TOMORROW'S ECONOMY

TOMORROW'S ECONOMY

A Guide to Creating Healthy Green Growth

PER ESPEN STOKNES

FOREWORD BY PAUL HAWKEN

The MIT Press
Cambridge, Massachusetts
London, England

This book was set in Adobe Garamond Pro by New Best-set Typesetters Ltd. Printed and bound in the United States of America.

Library of Congress Cataloging-in-Publication Data

Names: Stoknes, Per Espen, author.
Title: Tomorrow's economy : a guide to creating healthy green growth / Per Espen
    Stoknes; foreword by Paul Hawken.
Description: Cambridge, Massachusetts : MIT Press, [2021] | Includes bibliographical
    references and index.
Identifiers: LCCN 2020007225 | ISBN 9780262044851 (hardcover)
Subjects: LCSH: Economic development—Environmental aspects. | Sustainable
    development—Psychological aspects.
Classification: LCC HD75.6 .S748 2021 | DDC 338.9—dc23
LC record available at https://lccn.loc.gov/2020007225

10 9 8 7 6 5 4 3 2 1

To all who hunger for systems change,
the impatient ones

Contents

Foreword

If a writer wanted to sail into the most troubled and murkiest of academic waters, it would be economics. Although the economic profession considers itself a science, albeit a social science, it is not. It is descriptive theory about markets and value. Commerce is the root of economics. It is an ancient practice. We make, we buy, we sell. We use symbols of gain, precious metal coins or currencies, to engage in one of the oldest ways cultures connect— the exchange of value. The language around economics is imbued with water metaphors, making it seem as if modern-day exchanges are a natural phenomenon. Money flows like water. There is financial liquidity, we channel our investments, or, as happens in recessions, sources of capital dry up. Wealth trickles down, or, in good times, the rising tides raise all boats. These terms obscure the reality of economics as taught and practiced. It is not about the simple acts of commerce. Modern economics describes the complex financial dynamics of capitalism, an economic system that has accomplished three things. It has improved the living conditions of billions of people. Second, it has nearly destroyed the capacity of the Earth to support those conditions. Third, it has concentrated wealth in unimaginable ways, a profound imbalance of riches that disempowers and harms billions of people.

Tomorrow's Economy proposes a balance sheet of the world. In finance, a balance sheet discloses assets, liabilities, and capital. As Per Espen points out with both facts and flair, the balance sheet of the world leaves out all three accounts. What are the assets of planet Earth? Clean air and water, fertile soil, abundant fisheries, insects. We know only that they are all diminishing, but we have no total. And liabilities. The ocean receives 30 percent of anthropogenic carbon dioxide emissions in the form of carbonic acid. The acidifying ocean has reduced phytoplankton populations by 30 percent, yet phytoplankton are the source of 50 percent of atmospheric oxygen. And capital—what is our stock of natural capital? Pollinators, glaciers, rivers, aquifers, primary forests, wetlands. Economics does not account for these because it is based on the belief that there are no physical limits to economic growth as long as there is sufficient labor, capital, and technical progress. Financial balance sheets do not reflect the multiplicity of the Earth's resources unless they are commodified and sold. Economists are aware of the elision, albeit perversely. Nobel Prize–winning economist Robert Solow once admitted that if the biological limits of nature were taken into account, theories of unlimited economic growth would not be possible. Or to put it another way, there would be a scarcity of key elements for a healthy economy. Exactly.

Economists simulate reality using models. The scope of their models does not include tidal saltmarshes, gender equality, corruption, desertification, or burning forests—for starters. The models capture capital flows, labor markets, asset allocation, interest rates, inflation, but not life itself. How does one make sense of all this? To put it more simply, where do we go from here? How can we grow in prosperity and not burn the house down? How can we dramatically reduce our human impact while improving the lives of billions of people who deserve more?

Given the ecological collapse of species, systems, and regions, many now challenge the capitalist rationale, including a potent group known as ecological economists. On the corporate side there is the Davos paradigm of a future where the hyper-wealthy masters of capitalism consider adjustments to their financial momentum and accumulation by taking into account "externalities" including social injustice, loss of biodiversity, and global warming. There are entrepreneurs and consumer advocates who call for green growth, but exactly what that means is not well delineated. There are articulate cries for the complete cessation of economic growth as well as calls from degrowth advocates who would take it one step further.

Those who are aware and address the ravages of unimpeded economic growth want to rid the world entirely of capitalism. The venerable *Economist* dismisses such advocates as modern Luddites who would precipitate a backward leap to a stone age.

This is the rich foment of clashing views and beliefs that Per Espen has come to address. His training as both psychologist and economist gives him an unusual perspective. Virtually all economic models assume that economic behavior is predictable. Per Espen would demur on this point. Since his youth, he has been studying the intricacy and complexity of human behavior. Rather than disagreeing with conflicting economic theories, he pays close attention to all the voices. Perhaps his early training as a psychotherapist creates an openness and capacity to acknowledge the validity of different perspectives, and then, as he does here, propose a thoughtful, comprehensive way forward that takes into account inequity, poverty, ecological degradation, and capitalism—the need for more and the need for less. More prosperity for the needful; far less impact to our home planet.

The foundation of all commerce is value—what is valuable to whom, when, and in what way. The healthy growth approach you can find in *Tomorrow's Economy* is not a theory overlaid onto the existing economic tumble. It makes two important points. One, we cannot continue to steal the future, sell it in the present, and pretend it is GDP. That is what the balance sheet reveals when we do proper accounting. If the Earth were a company, this practice would be called fraud. Two, we can heal the future employing our youth, our entrepreneurship, our gumption, and our economic intelligence, and that would be an honest GDP. In short, we can either create assets for the future or take the assets of the future. One is called regeneration and the other despoliation. When we exploit the Earth, we exploit people and cause untold suffering. We are investing hundreds of billions in artificial intelligence to be applied by digital machines in myriad ways. However, we invest very little in the natural intelligence expressed by the living world. The biological world is exploding with new discoveries, compounds, and hitherto unrecognized intelligence. Plants have genetic memory going back millions of years. We now know that trees and plants have our five senses and fifteen more senses that we do not. Trees transmit a complex vocabulary through the air and roots to other trees. One of the most poignant and instructive discoveries is that forests are communities that physically and selflessly care for each other. Plants know what we seem to have forgotten, that the way we take care of ourselves is to take care of our community—communities of citizens and denizens who surround and uphold us, the animals, insects, fungi, and the trillions of bacteria in our body that outnumber our human cells. Although our innate human biodiversity has not been fully identified or quantified, we know that there are more genes in the human microbiome than stars in the universe, and approximately half are what are called "singletons," genes that are unique to each individual. Genes are intelligence. What are they saying?

Per Espen has been listening. It is his early training. We have been extracting value from forestlands, indigenous people, oceans, slaves, and the poor for centuries. You will find here a clear description of an economy that creates value for all, and for all whom we have diminished, one that does not impoverish any form of life, be it human, mammal, plant, fish, or creature. He honors and includes the skills, structures and brilliance that have evolved and brought us to where we are today, and pivots them away from the degeneration of the world to its regeneration. This is Per Espen's brilliance in this masterfully written and researched work.

Paul Hawken

Preface

When I was a teenager, the spine of a frayed yellow paperback caught my eye as I walked down the aisles of my school's library. *The Unpopular Series*, it read in Norwegian. Why, I wondered, would someone bother to write and publish something meant to be unpopular? I took it out and discovered it had a long essay called the "Present and Future—The Undiscovered Self" by psychologist Carl G. Jung.[1] At the time, I had no idea who Jung was. But for some reason my curiosity was piqued.

The essay, written in 1957, reflected on the state of the world after two terrible world wars. Both were started by Germany in a Europe that prided itself on having the best science, technology, philosophers, and historians of all time. Jung wanted to know how such a technically advanced culture, so famously built on "reason," could spark such atrocities. Rather than giving technical, socioeconomic, or historical explanations, he sought answers deep in the human psyche. How could the rational mind so easily succumb to its darker forces, emotional obsessions, and grandiose delusions? How could well-educated modern citizens, raised with the poetry of Goethe no less, be led astray by charismatic madmen who fired up the masses at rallies and used the media for fearmongering? What unconscious inner forces accounted for this present state of affairs, and what did they portend for our shared human future?

As Jung's questions and insights exploded inside my adolescent brain, I understood only half of what I read. Without even looking for it, I had found my calling. By the time I returned the book to the shelf, I had decided that I was going to be a psychologist, even though I had never met one and had little idea of what psychologists did. The small Norwegian city of Ålesund, where I grew up, was populated by engineers, fishermen, and small businesspeople—all proper materialists who viewed psychology with a deep suspicion that bordered on disdain. My family couldn't understand my choice: Psychologists were viewed as neurotic navel-gazers. My mother wondered what she had done wrong. Yet I knew I had found my path, so I kept my mouth mostly shut about my newfound direction and started seeking out ways to join what I now call the psych tribe.

Since then, most of my professional work—first as a psychologist and later as an economist, climate strategy researcher, and green-tech entrepreneur—has been related to understanding the future better, but always based on the present and always with a particular emphasis on how our dominant mental models determine our current actions and influence the way we imagine our futures. Like Jung, I find myself trying to make sense of a world that seems driven by unconscious dynamics toward self-destruction, despite our rational knowledge and technical expertise.

Some ten years after my library eureka, I finished my psychology training and set up a private practice in Kongsberg, another small Norwegian industrial town, mostly serving the weapons and high-tech maritime industries. Luckily for me, there were a lot of unhappy people around. Business was good. I had learned the lingo of the psych tribe. I had learned how to value empathy and emotion more than efficiency, and I could say hmm a hundred ways. Clinical work is deeply humbling: clients do not come with a user's manual. Each one brings a new worldview with them, and what's right for one doesn't work for the other.

Even so, a pattern emerged: Many were stuck in situations where the demands put on them by others overexploited their inner resources. Both at home and at work, only a tiny sliver of their fuller being was being appreciated and allowed to grow. Yet they had grown all they could in that one-sided way. It no longer gave them joy to perform that certain way at work, or to behave a certain way at home. Their bosses or parents, sometimes even themselves, would not accept any other way of growing, stunting their individuation.

After several years in the confines of the psychotherapist's office, I felt a mounting unease. Some patients were finding their way into a new flourishing, while others settled more deeply into their same old struggles. I was probably doing an alright job as a witness, a patient helper, a "non-judgmental judge" as one client dubbed me. But I started to feel stuck in the wrong place. I could see how our organizations, society, and economy were producing a steady flow of burned-out, depressed, anxious, and increasingly unhappy people who were overtaxing their inner resources in order to deliver ever more.[2] And I was stuck in a *reactive* mode—receiving those who had dropped out, been cast aside, or were otherwise resisting or unable to meet the demand to continually increase their labor productivity in our current economic machine. If I helped to fix them the way they or their company wanted them to be, to make them productive again, they would go straight back into the system that would eventually reproduce the same symptoms all over again.

Thus, I became disillusioned with the prescriptions from the sages of the psych tribe. Most were convinced that the inner life could exist, subjectively inside the mind, split off from the outer world. If only people could think more constructively, positively, or self-confidently, if only they could get their act together one by one, then society would get better, too, they reasoned. As if the soul can be healed while the biosphere and our institutions crumble, I thought. For by then, I had been awakened not just to the dire state of overexploited people but also to climate injustice, vanishing turtles and corals, and all the other realities of our overexploited ecosystems. I concluded that—as a book by cultural psychologist James Hillman had pointed out at the time, "We've had a hundred years of psychotherapy and the *world* is getting worse."[3]

There seemed no end to the steady flow of used-up, thrown-away laborers experiencing exhaustion, sleeplessness, worries over never being enough, unable to cope with expectations of family and work, and a host of fears instilled in them by a system with little or no insight into what "human capital" is really about. So, I switched focus to the economic machinery itself. It seemed better to proactively address the systemic cause—the way in which economic organizations choose to grow and treat their human resources—rather than trying just to treat the symptoms.

Disillusioned with the psych tribe, I turned to the econ tribe. I applied for entry by teaching organizational psychology at the Norwegian Business School. I've since spent many years there teaching executives (and soon-to-be executives) how to envision new strategic futures, specializing in futures thinking and scenario methods that, I hope, could build more respect for and inclusion of social and natural capitals into their strategy. I worked, as Jung's book had pointed out to me, with the "present and the future."

As I worked on laying out possible futures in economic terms, telling stories of the future in the present, I also became curious about how managers and economists actually think about the world. Why do the econs choose to describe the world with the terms they use? Why do they think the way they do about both human resources and natural resources? Why do they time and again exhibit severe limitations of imagination when it comes to innovation and disruptions?

Still fluent in psych lingo and aware of the power that language holds, I started looking at how certain root metaphors have shaped economic theories since the mid-1800s, solidifying into "truths" that were taken for granted in the 1900s. Core among these are the ideas and images of wealth as economic growth, starting with Adam Smith's *The Wealth of Nations*. Launching into a PhD, I learned econ's lingo, its favorite concepts and metaphors; I got to know its greatest forefathers and heroes (like Smith, Marshall, Keynes, Solow, Samuelson, Drucker, Porter, Stiglitz). In gradually becoming an econ tribe member, I also learned how to pray to and honor the holy demand-and-supply curves, how to tilt my head and say, "But what's that on the margin?" or "What about the alternative cost?" or "Have you considered the elasticities?" or "Remember the rebound effect" and much more econ mumbo jumbo.

However different the econ tribe is from the psych tribe, I discovered that their chiefs, too, turned out to have their own fixations, superstitions, rituals, and blind spots. For instance, the econs' capacity to imagine alternative futures is often severely restricted. Try as they might, they rarely move beyond a high, a low, and a middle business-as-usual equilibrium forecast to any given challenge. And that business-as-usual future inevitably prescribes one thing: endless economic growth.

The endless growth of the business-as-usual type has a dark side that's less easy to ignore with each passing year. Not only does it overburden our inner selves, but it also manifests dangerous realities like climate change and the breakdown of nature's life-supporting systems. With ever more power through technology, new energy sources, and strong economic growth, we are—without deep self-knowledge—transforming the entire planet Earth. Wrote Jung in 1955: "Coming generations will have to take account of [the unconscious man within us] if humanity is not to destroy itself through the might of its own technology and science" (§585).

Are our modern minds mature enough to understand what we are doing any more than we were in the early decades of the 1900s? Or is humanity playing around with Earth without self-understanding, once more yielding to charismatic strongmen promising to save "us" by crushing "them"? Will we again use the fantastic outcomes of technology, wealth, and innovations to succumb to the darker impulses of the ancient human psyche, accelerating our own self-destructiveness? Or will we be able to shift our inner and outer development in a new direction, onto a flourishing path?

Exploring those questions is where my two worlds, the disciplines of psychology and economics, collide and this book—an exploration of the mindset, mechanisms, and possibilities of a genuinely *healthy* growth—begins.

INTRODUCTION

What is the first thing that comes to your mind if I say *growth*?

The word evokes a multitude of meanings. "Growing up" implies getting taller, bigger, or older. Others point to improving the quality of something—as in "growing better." Some hint at power—such as "growing greater"—and still others refer to branching out or diversifying, such as "growing broader." Maybe you thought about volume: "growing larger," as with an inflating balloon. Or maybe you thought about spiritual growth—the inner, personal growth invisible to the eye. If you're a biologist, you might have envisioned sprouts, seeds, cyclic rebirth, and populations. If you're an investor, maybe asset growth (higher valuations leading to more money) came to mind. A darker version is found in cancerous growth—an excessive, unbalanced, and out-of-control growth.

Meanings, both quantitative and qualitative, abound. But when it comes to recent discussions about *economic* growth, something peculiar happens. Some see it in a positive light, as in "the country grew at a robust 3 percent last year." Others see it in a negative one, as in "consumption and economic growth must be stopped if we are to avoid global heating and unfair resource depletion." In recent public debates on economic growth, nuances have too often given way to either-or, pro-growth or anti-growth stances. This locks our thinking into a polarity: either expansion *or* limits, progress *or* regress, go *or* stop. Will further growth lead us to heaven *or* hell on Earth? All of which boils down to just two fundamental options in our collective mindset: growth is good, or growth is bad.

Such polarization leaves little room for reflecting on what *kinds* of economic growth we're really talking about, and what kinds of growth are worth fighting against or struggling for. This book aims to resolve that polarization by carefully examining the different ideas we hold of growth and distinguishing between the different models and directions of economic growth. It aims to battle with a controversial question: Is a healthy type of growth possible, and if so, what kind of therapy will it take to lure our economic minds away from destructive growth and toward healthy growth?

FIRST, THE BLESSINGS . . .

If you live in a wealthy country, it's easy to forget how incredibly well-off economic growth has made, decade after decade, vast numbers of humans. Most middle-class families—even if stressed or struggling—have higher living standards today than the dukes of medieval times. And never in modern history have so many humans been out of extreme poverty.[1] These are good reasons to celebrate economic growth. Material successes from running hot water to air travel, refrigerators, smartphones, antibiotics, anesthetics, cars, and wide-screens would have been miracles to people from the 1800s. Imagine having virtually all the information in the world available in a second, anywhere, anytime, out of thin air. Such successes have elevated and congealed the sanctity of economic growth in the modern mind. Threaten it by saying, "We must stop growth to save the planet!" and people will immediately defend growth by stating that we can't return to the Stone Age, with everyone living in mud huts.

Inside our brains, ideas of growth fire off neural sparks glowing with the golden hope of future betterment. Both economic growth and personal growth have become deeply and closely associated with positivity, success, optimism, strength, progress, and even invulnerability. Both inner and outer growth have been extolled. We—meaning all who participate in modern economies—seem to have programmed our brains for ever more of it. We have better stuff than ever before. And there's seemingly no end to it. United States politicians exclaim that they want to grow America to greatness. And others want to grow China to greatness, and the European Union, and India. The list goes on.

The world will no doubt need and see even more economic growth in many regions, not least to lift the one billion people subsisting on less than two dollars a day out of destitution. Humanity also needs more growth to settle another 1.5 billion people into rapidly mushrooming cities over the next ten to fifteen years, to generate better and more jobs everywhere, and for smarter mobility, nutrition, communication, energy, education, and healthcare, just to mention a few of the well-known yet amazing blessings that healthy economic growth can bestow.

Rising incomes will even lead to more open and democratic societies, argues economics professor Benjamin Friedman, pointing to evidence that when growth and trade thrive, stability and peace follow. Even cultural blessings like better morals, freedom of opportunity, tolerance, social mobility, fairness, and democracy have come easier with economic growth—while economic stagnation makes getting and maintaining them nearly impossible.[2]

If only we could have more growth, the prevailing logic goes, most things will be fine and good. Nobel Prize–winning economists such as Robert Solow, Paul Samuelson, Paul Romer, and Kenneth Arrow: high-profile economists such as Lawrence Summers; countless highly respected economics professors; and other growth missionaries seem incapable of or unwilling to recognize any physical limits to growth. When any kind of resources get scarce, they reason, their prices will rise, and then, lo and behold, market players will be incentivized to find new substitutes.[3]

The *Economist* has this to say about what the world's next great leap forward entails: "Nearly 1 billion people have been taken out of extreme poverty in 20 years. The world should aim to do the same again . . . Most of the credit must go to capitalism, free trade . . . and growth." And the *Telegraph*'s assistant editor Jeremy Warner states that "without economic growth, there's no prospect of the global solutions to climate change that are needed. You cannot force people to return to the stone age in the way they live. But with growth and money, there's now every prospect of bringing about the required transition."[4]

Such techno-optimist thinkers see a logic for how economic growth can go on endlessly, bringing ever more blessings to humans thanks to changes in market prices, substitution, new ideas, innovations, and human ingenuity. In short, with more growth, the future is bright! And they find it puzzling that others, mostly either tree huggers, lefties, or Luddites, have a problem with that.

. . . THEN THE CURSES

It's long been clear that not all kinds of growth are equally beneficial or healthy. All this merry whoopee for growth comes with noxious, shadowy undertones that threaten the celebratory mood from time to time. The way our modern economy grows is by most scientific measures actually sawing off the fertile branch that has nourished humanity for the last seven thousand years: destroying soils, acidifying the oceans, weirding Earth's climate, and killing off wildlife at breakneck speed. Also, most workers—full-time, underemployed, and certainly the unemployed—aren't invited to the party that the capital owners are enjoying. Inequality has skyrocketed since 1980.[5] To top it off, further economic growth no longer gives most people in richer countries any more satisfaction with life over time.[6] Hence, a majority of us miss out. In this view, economic growth is broken, corrupt, sick, and sickening.

Several prominent academics, activists, and anti-growthers insist on the physical impossibility of perpetual economic growth. "Steady growth is suicidal," says acclaimed Canadian professor David Suzuki, adding that the mindset that perpetuates it "is a form of brain damage." Writer George Monbiot calls growth "the destructive god that can never be appeased." *Guardian* economics correspondent Richard Partington writes that "Growth has become a holy grail for governments seeking re-election." Distinguished professor emeritus and energy expert Vaclav Smil concludes that "growth must end." And climate activist Greta Thunberg attacked all the grown-ups at the United Nations General Assembly: "We are in the beginning of a mass extinction, and all you can talk about is money and fairy tales of eternal economic growth. How dare you!" Anthropologist and degrowther Jason Hickel is clear that "growth is killing us." And bestselling author and activist Naomi Klein writes, "We have an economic system that fetishizes GDP growth above all else, regardless of the human or ecological consequences."[7]

Many have therefore recently gathered behind the degrowth banner, arguing for cutting consumption, particularly for the rich. They see no reason why more growth can solve the problems that economic growth itself has created. Rather, they call on everyone to sacrifice plane rides, cut plastics, and stop eating meat. They call for immediate, stringent sustainability regulations, carbon taxes, and a plan for prosperity without growth.[8] The wealthy and their money are ruining the world, they say. It's time to cut back, cut out, cut down in order to live together on a finite Earth.

It is as if the economic mindset that had such immense success directing the modern adventure after World War II has recently turned "mad, bad, and dangerous to know," as economics professor Steve Keen of Kingston University puts it.[9] As if it's now suffering some rigid psychopathology, an addiction to ever more of the same, an obsessive compulsion to repeat the same thinking, the same old patterns. In short, more economic growth clearly increases inequality and pollution, wrecks the climate, and no longer improves well-being or happiness. This logic leads anti-growthers to wonder why a clearly destructive system remains dominant. They find it puzzling—and frustrating—that conventional economists have a problem recognizing that with continual growth, the future is dismal.

CONFRONTING OUR INNER GROWTH PARADOX

As time shoves us relentlessly deeper into the twenty-first century and toward the once-distant climate thresholds, how will we choose to face the challenge? More growth or degrowth? It seems—from this debate—that we can't live without growth and we can't live with it. Damned if you do. Damned if you don't.

We *must* have growth. Yet further growth is impossible. A double bind.

This paradox is not just theoretical. It shows up in the groups of people that I work with in very different ways, depending on their inner imaginings of growth. On the one hand, I speak with mainstream managers and MBA students, petroleum engineers, unions, financial investors and brokers, energy economists, and politicians. But on the other, I collaborate closely with environmentalists, Green Party politicians, localists, slow-growthers, degrowthers, climate researchers, and animal-rights and social-justice activists.

The first set of people can't get enough of growth. The second group struggle to even mouth the words "growth" or "green growth" in a positive way. Personally, trying to build a bridge, I keep falling into the chasm. Among mainstream economists and growth investors I'm viewed as too green, too closely associated with emotional idealism and activism to be balanced and objective. Among environmental activists I'm seen as too business-friendly, keeping company as I do with those who speak of further growth. Attempting to reach out to both camps at once, I've frustrated every-one. My colleagues and coworkers at the business schools are deeply focused on issues like marketing, globalization, strategy, change management, digitalization, accounting, and law. They struggle to grasp what "green" or "sustainability" has to do with their subject. That's something for the Environmental Protection Agency and politicians. Yes, the United Nations' Sustainable Development Goals (SDGs) are nice, and indexes of corporate responsibility sound good, but the real beef is in the earnings. Face it, they say, if a company goes into the red, it doesn't matter how green it is. You're out. Bust. Bankrupt. Gone. So, grow or die! Black is still the new black.

Then there are my colleagues and friends in the sustainability movement, who strive for a transition to a greener, more just world. They are rebelling, taking to the streets, fighting a losing battle to save the last shrinking remains of corals, rainforests, and glaciers. Their voices fill with contempt whenever I speak of growth, gross domestic product (GDP), airplanes, the petroleum industry, corporate profits, or anything else to do with our current economy. The eyes of new acquaintances from this camp glaze over if the first thing they learn is that I work at a business school or as a paid consultant to big industry. The economy is the enemy. Powerful. Dangerous. Destructive. Unethical. It caused the problem and is the problem.

The pushback has been discouraging at times. Maybe I should call it the *green squeeze*: The greens say I sound like an economist (that's an insult, I presume). The econs say I talk like a green hipster (an insult, too). Yet mediation and problem-solving are my calling. So, I listen to both sides and try to find new directions toward some common ground. That's why I lecture at the Norwegian Business School in Oslo, run the BI Centre for Sustainability and Energy, and do strategy consulting for businesses. It is also why I participate in politics as a Green Party member, publicly debating the future of oil industry subsidies and assisting the environmental NGOs.

This book is my personal answer to both camps, and to any reader who is struggling with the issue of growth, both personally and theoretically, and is curious about the emerging framework for economic thought in the twenty-first century.

It would be false and self-defeating to feign impartiality when trying to establish common ground between disparate groups. I'm deeply critical of conventional economic growth, originating as it does from coal mines, slavery, colonialism, fossil oil combustion, worker exploitation, and ecological destruction. Yet just "saying no" to economic growth is not a viable way onward. I am, however, convinced that a better version of growth is possible—a healthy growth that can replace and repair the faults of the old gray growth model.

A HEALTHY VERSION OF GROWTH?

The strongest driving force behind both the successes and excesses of economic growth is the hard-coded manner in which the economic system expands. It grows in uneven ups and sudden downs. Yet over time a capitalist economy can't help but grow. Growth is built into its DNA. Machines beget more machines. Capital begets more capital. More production gives more profits, which in turn are reinvested in even more production as the economy gets wealthier. Our accounting systems make sure this is happening and calculate return on investments mechanically and accurately (at least accurately in a mathematical, if not a real social or environmental, sense). The purpose is to grow more profits, of course. It is what systems thinkers call a self-reinforcing loop. This is a causal explanation of why there is seemingly constant economic growth. Money generates more money, particularly for those who already have it. That's the core engine of any version of the many capitalisms out there.

The negative side effects of continued conventional economic growth are by now more than obvious. Critiques are a dime a dozen. Only a few of them are mentioned above. Yes, there is no lack of touted alternatives. There is much fanfare around such concepts as low-carbon, circular, sharing, slow, local, solar, socially responsible, inclusive, 100 percent renewable, regenerative, and sustainable economies. Indeed, there are many, many sustainability initiatives under way.

Yet beneath all these remedies, opportunities, and solutions, a deep disillusionment and questioning of further growth itself lingers in the mind. Are all these attempts to modify the current growth model feasible, viable, or truly *sufficient*? Isn't all this "sustainability" and "circular" buzz really just hype and sugarcoating on the outsides of an inherently unstable and destructive core in the current economic system? Isn't everyone who participates in the modern economy also supporting the financialization of the world—turning everything into money for consumption? Aren't most executives still championing an overall system that shortsightedly focuses tremendous force and attention on bonuses, short-term whims and wishes of consumers, and investors, while at the same time busily sawing off that same branch we're all sitting on? What's the point of struggling to win a few cosmetic changes on the surface (like corporate social responsibility reports, rooftop solar panels, or a meat-free Monday) if the deep core continues its unperturbed destructive swirl?

This doubt has good grounds. So-called greenwashing of the gray growth model has been widespread. Greenwashing (and the newer "green-wishing," "impact-washing," and even "SDG-washing") means that politicians and corporations highlight a few green actions here and there but avoid addressing inaction on the most substantial issues: their contribution to the worsening megatrends of inequality, global heating, and the demise of oceans and wildlife.[10] Covering over those shortcomings with peppy speeches and glossy ads showing photos of hands holding green sprouts, gives sustainability endeavors a bad reputation. In short, greenwashing results when communication and marketing are stronger than delivery. There is talk but little walk.

Amid all the greenwashing, the idea that growth could be radically revised to reverse such megatrends may seem too good to be true. Despite being told that there may be a healthy version of growth, many remain deeply suspicious. Conceivably, some *small* businesses, their argument goes, like an organic slow food restaurant connected to nearby farms, can have healthy growth for a while, but it probably can't scale up. Healthy growth seems doubtful, they say, at city, state, corporate, national, or international levels. The minute you've gone really big, with listed stock and revenues in the billions, the likelihood for continuing to achieve healthy growth plummets. You're now run by the stock markets. No longer serving the local farm but ruled by the global financial system, the greedy capitalist class, the selfish 1 percent suffering from severe affluenza.

Does this explain why believing in sustainability often feels like being complicit in some larger hoax? Is even sustainability corrupt? Maybe it's time to raze green growth and even "kill sustainability."[11] Your inner critic may whisper that it won't work anyway: "If you still believe in the possibility of creating true sustainability by tinkering at the fringes of the economy, then you're just a fraud."

Such doubt and dissonance are perhaps lingering somewhere deep in your own mind. You might be asking yourself if your participation in continued corporate and market growth will cause the future world to have fewer wonders. Will everything wild be gone anyway in hothouse Earth? What happens if we really don't succeed in changing the DNA of economic growth? Won't further growth, even if labeled "sustainable," "green," "decoupled" or "healthy," cause more inequality, social tension, and harm to oceans, wildlife, and climate? This is the kind of soul-searching under way among even leaders in the sustainability movement.[12] There is also widespread confusion about how to go from here in a *credible* way that convincingly adds up from micro to macro, as witnessed by the proliferation of definitions, rankings, metrics, and indexes of sustainability. Would it be better to withdraw completely from the economic machine, retreat into some patch of remaining wilderness or community, and fight any intrusion 'til the bitter end?

The pro-growth camp insists that further growth and better technology can take us through the current crisis, particularly if governments price the externalities (the pollution) properly. On paper this is simple: The polluter-pays principle addresses the current market failures of resource overuse and climate change. Politicians and voters ought to listen to the economists to get the prices right.

The anti-growth camp insists the whole market system itself needs immediate transformation, and to achieve that, growth must be stopped (at least in the rich countries). Consumption should be reduced, and the corporate takeover of society and politics must be reversed. And the rich capitalists thoroughly taxed.

Based on this debate, it's easy to imagine that a "third position"—as promised by this book—would be to seek middle ground. After all, I accept that both camps have a point. But my argument is not to seek a middle ground or a compromise. Rather, I suggest we accelerate growth as much as we can, but in *a new direction*. And that we reframe what it is that can and should grow. To do that we (any changemaker) can utilize and supercharge certain trends and innovations already visible in the economic system. Innovations and market drivers are available now that make rapid healthy growth possible—but only if transformations in government practices and individual behavior come along with them. Changemakers can push for and verify whether every part and player in the economic system works and moves at a sufficient speed in this new direction to create profits while at least doing their fair share of the transition.

I myself am a growth critic. I'm deeply skeptical of the old economic growth model and theory that dominated the twentieth century and still persists. So I don't embrace growth as we know it. I'm similarly wary of the organized corporate forces that will strive to derail, belittle, or capture any real progress toward genuinely healthy growth. But I prefer to express my criticism not by the all-too-easy verbal bashing of growth, corporations, and capitalists. Rather, I will try to put forth a constructive, realistic alternative building on data, trends, research, and solutions already available to start healing growth's self-destructive sides.

The main question driving me and this book's exploration is this: *Can a healthy growth model outcompete and replace the gray growth system in time to keep within Earth's safe operating space?*

SYSTEMS CHANGE THROUGH HEALTHY GROWTH

As millions of youths filled the streets in 2019, inspired by Greta Thunberg's climate strikes, a common banner read "System change, not climate change." How do these kids feel? They express a sense of betrayal by the previous generation—the boomers—and the economic system they created. This old system is giving kids a sense of despair over the burning, boiling, flooding, unfair state of the world that they are inheriting. From marching with them in the streets and speaking with them, I imagine their feelings are akin to what the psychiatrist Robert Lifton describes as "an amorphous but greatly troubling sense that something has gone wrong with our relationship to nature, something that may undermine its capacity to sustain life."[13] In 2019 it was no longer a case of stating that "we should think" about the consequences of emissions on the future generations, as was commonly said back in the 1990s and early 2000s. Instead, that future generation is here, now, demanding immediate system change.

But what exactly *is* system change—at levels from local companies to countries? How can you tell if the system—or a part of it—is changing each year or not? If it's on the right track or the wrong track? Too slow or rapid enough change? And most important: What solutions will bring about the wanted system change in the most effective way?

Here's what I propose to all changemakers: to create genuinely healthy growth models in all market economies that rapidly outcompete and replace the gray growth we know too well from the twentieth century. We need to reimagine and spread new models for value creation for the twenty-first century: a credible, feasible healthy growth approach that can reconcile the two camps. Simply put, economic growth is healthy when the value creation from any economic actor gets sufficiently more resource smart and more inclusive each year. The vision is to keep growing human well-being while fitting the footprint of the economy inside Earth's safe boundaries. We (in the sense of anyone engaged) need to be able to monitor those annual changes to keep decision makers accountable and avoid greenwashing. Healthy growth is measurably profitable, more resource productive, and more redistributive by design each year. It adds up from micro to macro and impacts troubling megatrends fast enough to matter.

Just like economics, this book has its micro and macro planes. It will distinguish between healthy and unhealthy economic growth at personal, corporate, city, national, and global levels. You can find tools for the nitty-gritty practice of stepping away from destructive growth and encouraging healthy growth.

As you'll see in the pages ahead, it is possible to move beyond the current obstacles and break new ground. Fresh ideas, new practices, and deep driving forces that are already at work in today's markets can accelerate a transformative approach to growth in the coming decades. We will explore how to move from "gross growth" (as in bigly, grossly, gross domestic product) to "healthy growth" (as in balanced, regenerative, equitable, and long-term wealth accumulation). The quest involves going beyond the thinking in conventional enlightenment economics from the 1900s and expanding on the parts still fit for the 2000s. It entails looking for what can be scaled up to create an "enlivenment economics." It implies a shift from narrow, calculating self-interest to a broader, relational, and inclusive Self-interest (with capital S) that delights in enlivening the network of Others flourishing around us.[14]

My wildly audacious ambition is to welcome economic growth into the room as a patient personified and explore a psychotherapy for it. My theory of change is to help growth rethink and reframe itself over the coming decade.

WHAT'S NOVEL IN THE HEALTHY GROWTH APPROACH?

If anything, the sustainable growth topic is old hat. There are tons of books about it, at least since 1972.[15] I know. In a way it's crazy to write yet another one. But where do we stand in the 2020s? And why do I believe that healthy growth is both possible and *different* from what you've heard about before? This book both builds on and synthesizes the current leading knowledge, so that experienced sustainability practitioners will find a lot of common ground and references in the endnotes. The healthy growth approach is well grounded in the economic and sustainability research of the previous decades, and I think it's important to acknowledge that foundation.

Throughout, though, the healthy growth approach brings together in new ways the recent research findings from three separate domains: cutting-edge economic psychology, which probes the mindsets of economic growth and inequality; innovation economics, which studies how new solutions arise and spread (or not) in market economies; and Earth system science, which is giving us science-based targets for resource use and environmental footprint. A new understanding of economic growth can be found in the integration of these domains. I want to tell not a science fiction but a plausible story of how the coming decades could unfold into deep systems change, as ever more actors start to do what is needed each year.

But each part also goes *beyond* conventional sustainability thinking in major, unique ways. Part I brings in wholly new and refreshing perspectives on the psychology of economic growth to show how to move beyond the pro- vs. anti-growth stances. This part examines the archetypal psychology and the stories that drive the imaginative changemakers of all main sectors. It also describes the socioeconomic drivers of a sixth wave of innovations that have been mostly undercover but are taking off big time during the 2020s. The storytelling applied throughout helps with restor(y)ing growth and enlivening the economic concepts and theory.

Part II introduces the novel healthy growth compass with detailed metrics to help navigate growth in the twenty-first century. The healthy growth compass lets you "see" whether each dollar of value creation aligns with a lower environmental footprint and a more equitable society. This is a more in-depth technical section that explores how to improve productive, social, and natural capitals in a balanced, beneficial way. The healthy growth framework is woven around a pro-business and national economics core concern: expanding value creation. By consistently focusing on value added (gross profits) in integrated metrics, business interests can align operations with environmental and equitable concerns in a systemic, self-reinforcing way. We can reframe the old GDP numbers in completely new ways—for instance, to check whether any growth is both sufficiently resource productive and inclusive. Both pro- and anti-growthers, both long-term investors and activists should be able to embrace growth in value added that brings sufficient resource productivity and social productivity.

Healthy growth also comes with a new logic for built-in climate justice; when richer countries decarbonize sufficiently in absolute terms, it gives developing countries space to grow. Yet the same fair rates of change apply to everyone (chapter 7). Similarly, it lets both larger and smaller companies and cities with higher emissions calculate the targets and performance for them to do their fair share of the transformation. An audited healthy value creation that scales from micro to macro is needed throughout the economy, at the company, city, county, sector, and country level. The science-based targets and objective approaches presented show how to avoid the greenwashing that so often hampers attempts at sustainability.

Part III gives tools and strategies for how to accelerate the system change to cleaner, safer, and healthier growth in practice. It starts by outlining the triangular system of citizens-business-government and explains why all three corners of this triangle have to move in sync to succeed in the transition to a healthy economy for all. Whether you're just an individual (chapter 9), part of a company or an NGO (chapter 8), or an engaged citizen with a concern for governance of the markets and the commons (chapter 10), you can find concrete examples and actions that help speed up the transition to healthy growth. Part III ends with the surprising explanations on why both population and GDP growth will slow down over time with higher GDP per person during the twenty-first century. Well-being, however, can keep growing. The final chapter (11) provides four science-based answers to the question of how it all adds up.

In short, part I lays out the helpful thought patterns and stories to resolve the growth paradox and give a sense of grounded hope, while part II provides us with a map of the problematic territory along with a compass for navigating to a better destination. Part III provides the skills, leverage points, and strategies to work on a healthier growth as we're walking, wherever we are.

A final difference from other economic books, is that I emphasize thought patterns, story, metaphors, feelings, language, framings, and visualization of complex relations. All these "softer" psychological characteristics are often overlooked in the more rational rhetoric of economics.[16] But this is how I integrate my psychological approach with (ecological) economics: not just the numbers and the facts, but how we think and feel about them too.

WHEN WILL WE SEE THE BROAD ECONOMIC SHIFT TO HEALTHY GROWTH?

When can we expect the new growth model to have fully outcompeted and displaced the twentieth-century model? Can it realistically happen in time to prevent runaway eco disaster with the associated breakdown of society, or runaway social disasters with the associated breakdown of ecosystems? Is our current predicament so bad that it's necessary to fool ourselves a little in order to keep up the faith necessary to go forward with enthusiastic action? The final chapter (11) offers my take on these grand questions, based on a number of recent modeling efforts looking at combined Earth and economic systems.[17]

The race is on to deliver a new economic operating system for the twenty-first century that is both compatible and conducive to nine billion people flourishing within one Earth. If you agree it's time for a mental upgrade, fresh ideas, and new stories of what healthy growth looks like, read on.

I RESTOR(Y)ING GROWTH

1 YOUR BRAIN ON GROWTH

As a child grows up, she gets stronger, taller, bigger, and more independent, resilient, and mature. Yet to grow up is so natural to us that we rarely if ever reflect on the concept itself.

The two small, seemingly innocent words *grow* and *up*, however, make a lot of difference to human beings. They connect to something very fundamental: Each and every one of us started out as a small toddler, then increased—first in height and weight, then in capacity, power, and influence. We know deep inside that growing brings us closer to mastering the world around us, expanding our impact. In our brains, the neural networks that fire when we hear *grow* spark a web of associations with strength and health, and they have done so since childhood, when we were at our most sensitive and malleable.

So, grow leads to up, and up is good and happy. As the cognitive linguists George Lakoff and Mark Johnson wrote in their now classic study *Metaphors We Live By*, we say "I'm feeling *up*. That *boosted* my spirits. You're in *high* spirits." But there is a converse: "He . . . *sank* into coma. He came *down* with the flu. His health is *declining*."[1] These up-down/growth-decline associations are utterly obvious when you start to reflect on them, but less obvious is how they subtly shape our language and worldviews. The most fundamental values in our culture align with our favorite metaphors— and these embodied, generative metaphors shape how we think, feel, and communicate.[2] At our most basic neural level, grow = up = good = survive, and fall = down = bad = die. This framing affects how we view events in our own lives. And it spills over into economics and politics, too.

Our fascination with and enthusiasm for growth of all kinds stems from deep roots in body and language: Like trees from seedlings, we humans grow from humble beginnings. We attempt to stand tall in the world. Compete to find our place in the sun. The same ideals are applied to society: When cities, business, and societies grow, we imagine them stronger and better. More money and more goods often add up to feeling excellent and being on the rise into the light. The future will be up, too, as long as we can have more growth. These deep-seated embodied metaphors shine through in the way most politicians and economists speak about economic growth.

THE PREVAILING MINDSET OF THE GROWTH PARADIGM

In 2008 the World Bank summoned a group of seventeen political leaders and two academics—Nobel Prize recipients Robert Solow and Michael Spence—to report on the state of growth worldwide. "Since 1950," they wrote, "13 economies have grown at an average rate of 7 percent a year or more for 25 years or longer. At that pace of expansion, an economy almost doubles in size every decade." Their report, they continued, focused on "sustained, high growth of this kind: its causes, consequences, and internal dynamics. One might call it a report on 'economic miracles,' except that we believe the term is a misnomer. Unlike miracles, sustained, high growth can be explained and, we hope, repeated."[3]

So, a repeatable miracle for getting a sustained *high*: That's how these luminaries of economics and governance view economic growth. The study team's language reveals their ideal state, their guiding star: They prefer growth rates "of 7% or more." From the psychological perspective, where we listen for the emotion underlying the reasoning, their language also reveals a disappointment at lower growth rates as well as a worry: that growth might fall again too soon. And a longing: for a growth miracle, but preferably one they can explain and control.

Such is the prevailing mindset created and upheld by the huge-growth discourse.[4] Another Nobel-winning luminary, Paul Samuelson, who wrote the bestselling textbook in the history of economics, stresses that the US economy grew eighteen times bigger from 1900 to 2000. He called this the most important economic fact of the entire century. And it's clear that he felt thoroughly, patriotically impressed by this growth rate, more so than anything else from the 1900s. To substantiate his argument, he reminded his readers that continued economic growth makes it possible for developed industrial countries to "give their inhabitants more of everything: better food, larger houses, better health services, pollution control, education and pensions. Countries that are successful in the competition for economic growth, as the UK in the nineteenth century and the US in the twentieth century, continue to climb in the international pecking order."[5] They also become role models for other countries that are climbing up to greater wealth—a climb that China, for instance, has decided to take on over the last decades.[6]

"Economic growth is clearly the most important factor in the long-term success of nations," Samuelson concluded, falling in line with conclusions drawn by Solow, whose seminal work on growth started in the 1950s. Since then, the assumption has spread in economics that growth is a continuous process that will persist forever, thanks to improving productivity, particularly from technical progress.[7]

A closer look at the language of political economics can be revealing: After the Great Recession of 2008–2009, politicians started talking about economic *recovery*. Calling the next phase a "recovery" implies that growth has fallen off track, been sick, gone *down*. And health is to be recovered through vigorous growth rates, the ticket to rising back *up*. Years later, in 2017, a *US News & World Report* economic analyst described the recovery this way (emphasis added):

> After 18 months of *slogging* through the single worst economic *down*turn since the Great *De*pression, the US in June 2009 finally *limped* out of the Great Recession and began what has simultaneously become one of the longest and most *disappointing* recoveries in modern history. . . . Average real GDP growth during the ongoing expansion has registered *just* 2.1 percent annually. Between the 1970s and the 1990s, average growth maintained a 3.3 percent pace.[8]

Even a recovery that had lasted for more than eight years is disappointing to this analyst, because the growth rate hadn't risen quickly enough, or high enough. And when there's not enough good, there must be bad: The piece's title, "How Long Will the Recovery Last?," evokes the specter of an eventual *fall*, or downturn—something that triggers a fear that all we have will be lost. Our miracle will cease and desist.

We hear about that miracle time and time again. As the *Economist* once put it,

> over the past century or so . . . the western industrial democracies have experienced what can only be described as an economic miracle. Living standards and the quality of life have risen at a pace, and to a level, that would have been impossible to imagine in earlier times. . . . All this has been bestowed not just on an elite, but on the broad mass of people. In the West today the poor live better lives than all but the nobility enjoyed throughout the course of modern history before capitalism.[9]

Growth in this context is not just a heavenly miracle but also a real-life salvation—and nothing short of hell awaits if growth slips away, opening us up to poverty, mass unemployment, loss, and social breakdown. In other words, when growth is seen as an absolute lifeline, fear becomes its cultural alter ego. It's no surprise, then, that political rhetoric—that messaging that plays on our hopes and fears alike—uses this reality to its great advantage, further locking us in to the growth-is-good, big-growth-is-better, no-growth-is-scary mindset. This is the mindset of addiction. It reinforces a widespread and deep-seated psychological addiction to growth. How long will the bottle last? We worry about the next shot.

Most investors and politicians addictively demand or promise economic growth annually, forever. For this widespread growth mindset, 2 percent annual GDP growth is not enough. We must get 4 percent or more, as Donald Trump has opined.[10] At the time, Pew Research had found that six in ten people in the US considered the economy, which grew by 1.5 percent in 2016, to be "very or somewhat good." Reported Pew, "This is the most upbeat assessment of US economic conditions . . . since 2007."[11] But bigger is better. Even if people don't really know what economic growth is in statistical, accounting, or academic terms, they still really want it. Lots of it. The desire seems to tap directly into some deep pleasure spot in the human brain. Hence, in poll after poll US citizens consistently rank "the economy" as one of their top priorities.[12] It seems that this economy is imagined, simplistically, to be either big and bullish or low and bearish. And circumstances in their own lives—like stagnant pay, joblessness, or massive student or medical debt, even in the face of overall growth—can be overlooked if a miracle awaits.

Political leaders and the public they influence have for a very long time acted as if gross economic growth is an overwhelmingly good thing. This is the long-standing production of the mental growth paradigm in Western cultures.[13] Thus, our brains have become programmed to hear about growth in glowing terms. Seeing graphs with rising curves, we expect salaries to go up, and lives to improve. Something is deeply wrong if a company or country is no longer growing. Yet a different reality is slowly settling in. All these numbers going up *used* to be a good thing, but over the last decades more and more observers have stopped singing the praises of economic growth. These discordant voices in the economic choir are telling us to look harder at the way rising growth curves have equated with rising inequality, rising carbon emissions, rising resource depletion, and, for many, a stagnating or deteriorating well-being. The conventional growth model that once looked like a lifeline, they say, is really a shiny but dangerous lure.[14] The more we hear about this downside of up, the more our ideas of unfettered growth take on a sinister tinge.

THE GROWTH OF THE GROWTH PARADOX

The second half of the twentieth century was, perhaps, the heyday of our giddiness over economic growth. In the golden 1960s, the United States experienced over 5 percent annual growth. The middle class was flourishing, with money to buy homes, cars, vacations. Which is why, when a group of computer scientists and systems thinkers at MIT published their *Limits to Growth* study in 1972, all hell broke loose.[15] The book was full of abstract graphs, numbers, and tables. It built on an abstruse quantitative model called World3, built around a spaghetti diagram that was utterly unintelligible to most onlookers. And in spite of all this it rapidly became a global bestseller. It sold nine million copies and was translated to twenty-eight languages. It has been updated twice, and its original business-as-usual scenario has—despite all the criticism—unfolded pretty close to reality so far.[16] Why all the fuss over a book that relatively few people actually bothered to read, and certainly—judging by the rampant mischaracterizations of its content—not all the way through?

One reason was simple: It stuck a pin in the balloon of the high-growth believers. It projected that conducting business as usual, with unfettered industrial growth as a guiding light, would leave us in perilous shape some decades into the 2000s, when drastically rising populations, consumption levels, and associated resource use would overshoot the planet's resource boundaries. We could rethink growth, the MIT team wrote, or plan on experiencing devastating levels of pollution, resource scarcity, social unrest, and wildlife loss during the twenty-first century. In other words, the planet itself was our true lifeline, and unchecked growth was actively destroying it. It was a dire warning.

But the believers' balloon still didn't deflate. To reflate it, growth economists and pundits ridiculed *The Limits to Growth* within a week of its publication. Yale economist Henry Wallich dismissed the book in *Newsweek* as "a piece of irresponsible nonsense," since it dared to question the notion of endless growth itself.[17] The discussion even reached presidential levels, as Ronald Reagan repeatedly declared, "There are no great limits to growth . . . when men and women are free to follow their dreams . . . because there are no limits of human intelligence, imagination, and wonder."[18]

Reagan was right in one respect: imagination and wonder are no doubt endless in their expanse. They are as infinite as they are intangible. Tangible stuffs, however—like food, energy, real estate, forests, wildlife, and soil— are unavoidably finite. We've seen, for instance, a 60 percent decline in wildlife populations since the 1970s as ever more natural areas have been claimed to feed our growth.[19] A group of researchers at the Stockholm Resilience Centre, led by Professor Johan Rockström, now head of Potsdam Institute for Climate Impact Research and one of the world leading climate scientists, dubbed the period since 1950 "The Great Acceleration." During that time, human population, foreign direct investment, urban population, energy use, great floods, fertilizer use, transportation, tourism, and much more have grown exponentially.[20] Suddenly, growing up and up has triggered fears of its opposite: threats of decline and breakdown. Maybe the clearest and most acute physical limits yet are not lack of virgin material resources but the capacity of Earth's sinks. Those are the limits of the air to receive CO_2 without abrupt warming, and water's capacity to absorb excess pollution, particularly nitrogen from fertilizer.

These concerns have become mainstream, entering our living rooms, boardrooms, and classrooms, write authors from the global consulting group McKinsey & Company.[21] Skepticism about the current growth model is now spurred by public leaders from Pope Francis to former UN Secretary-General Ban Ki-moon, and from Unilever's Paul Polman to BlackRock chief banker Larry Fink, who wrote in a letter to all Standard & Poor's 500 top companies that "Short-term thinking pervades our most important institutions, from government to households. We've created a gambling culture in which we tune out everything except the most immediate outcomes."[22]

Despite this emerging mainstream skepticism, the limits-to-growth debate remains, as it has been for decades, mostly stuck in the ditches between the politically dominant pro-growthers and the vocal anti-growthers. While one side champions the benefits and durability of growth, the other repeats the impossibility and destructiveness of continued growth on a full planet. Is there something missing in the debate?

According to Professor Jorgen Randers, coauthor of the original *Limits to Growth* study and my colleague at the Norwegian Business School, thirty to forty years of public debate have been wasted by ignoring the conceptual murkiness and emotional muddle in people's perception of growth. In hindsight, he says, the analytical solution is simple. Maybe the warnings could have been more effective if they had more clearly distinguished between the two main markers of growth: intangible money on the one hand, and physical materials on the other. It is possible to have continuing growth measured in money (which is now really just immaterial, intangible numbers stored electronically somewhere in the cloud), but one can never have infinite growth in material resource use nor its associated ecological footprint.[23]

Thus, there are clearly material limits but not necessarily monetary limits to economic growth, just as there are no limits to imagination. The logical solution, then, is to require that all future monetary growth decouple fast enough from material resource use. And to monitor and measure both sides of growth to ensure that it happens at sufficient rates of change to avoid ecological deterioration. This means that for each extra dollar of value creation, resource productivity must grow even more quickly (see chapter 5). On the mathematical level, this distinction between monetary and material growth seems clear enough. But in everyday language and communications, full of emotion as politics are, "growth" takes on other psychological associations, symbolic markers, and values.

Ideas such as circular material flows, decoupling, and resource productivity are indeed creeping into transformative business models, promising further growth opportunities while resource use goes down. But there remains a huge, emotionally charged polarization between growth worshippers, who deem any type of growth simply splendid, and those who abhor any kind of economic growth, regarding it as a ticket to the apocalypse. These positions often stem from preexisting psychological attitudes.[24] They emerge from the narrative one holds of growth and one's own self and wealth, as much as from rational, quantitative, or economic reasoning. People tend to first adopt a viewpoint—based on their long-standing values, habits, wealth, and lifestyle—and only afterward find and refine arguments that bolster that viewpoint. The psych tribe has words for this all-too-human tendency: confirmation bias, motivated reasoning, identity protection.[25] In other words, we first gravitate toward our comfort zone and then use our brains to defend it.

So, the debate on growth still rests in a rather peculiar state. First, there is the physical impossibility of continuing with conventional, gray growth. Anyone who has acknowledged, for instance, the enormous amounts of plastics entering our oceans, followed the disappearance of tropical forests and glaciers, or has had even a superficial exposure to Earth system science can see that yet more resource use and waste, along the conventional twentieth-century growth model, will bring climate disruptions, kill off even more wildlife, pollute the fresh waters and the oceans, and destroy ever more of the remaining topsoil.

But then, the opposite of growth—degrowth or contraction—also seems impossible. If economies start collapsing (with GDP declining), then unemployment shoots up. Both investors and people hit the brake pedal on investing and consuming, which drags the economy down further. As more people lose their jobs, poverty and inequality threaten the social fabric. Homes are foreclosed. Debt problems become insurmountable. Creditors and governments call for austerity measures. The lifeblood that keeps our modern society pulsing—the flow of money through the banking sector—threatens slow to a trickle. People stop noticing and prioritizing nature.[26] Long-term concerns are put off in order to deal with the immediate threats. Trust in ineffectual politicians and government declines as people suffer from the fallout. Populism blooms. Also, new innovations spread dramatically slower, if at all, unless supported by fresh government funds—which are dwindling. The transition to clean power and investments in resource efficiency lose momentum. Innovative startups and scale-ups lose funding as investors want to avoid risk and keep money safe. There is little or no money available for infrastructure maintenance, cleaning up old pollution, or reinvesting in natural capital. So degrowth (in the sense of GDP decline[27]) also seems impossible or unwanted.

What about green growth, a concept that grew to prominence and popularity during the 2000s? The World Bank defines green growth as being "efficient in its use of natural resources, clean in that it minimizes pollution and environmental impacts, and resilient in that it accounts for natural hazards."[28] The European Commission writes, "The aim is to create more value while using fewer resources, and substituting them with more environmentally favorable choices wherever possible."[29] But many critics claim that green growth rhetoric just aims at better efficiency and somewhat more sustainable consumption while disregarding ecological limits. Consumption growth soon eats up the efficiency gains. Rebound effects can make resource use boomerang back up.[30] Green growth becomes, at least in practice, mostly a continuation of the conventional economic growth model, just under a new label. With this, disillusionment breeds.[31] Green growth begins to sound like "clean coal." So, the greenwashing of growth doesn't help either.

In summary, then, those meandering into the growth debate find it difficult to find a credible way forward. Conventional growth is impossible. Degrowth is impossible. And green growth is often just greenwashing. In a world taken over by global corporatism, it is simply too difficult to imagine that social and environmental considerations will be anything but swept aside the minute they threaten profits. No wonder many are left with the perception that the whole conundrum is unsolvable.

Others have just given up on the idea of growth itself, and decided to call it something else. Some thinkers and authors have suggested "development." Then, we'd no longer have growth as we move into the twenty-first century. We'd only have development—in the same way that ecosystems don't get bigger, they develop into more nuanced patterns and more diversity. Others, like ecological economist Tim Jackson, have gone with "prosperity without growth," while Peter Victor suggests "managing without growth," and hence supporting post-growth initiatives.[32] Others, like author and physicist Fritjof Capra, ask us to focus on qualitative growth, not quantitative growth.[33]

But calling for development to replace growth is simply to repeat the trajectory of *Our Common Future*, a famous 1987 report by a UN commission led by former Norwegian Prime Minister Gro H. Brundtland. This report coined the term "sustainable development," which its authors defined as "meeting the needs of the present without compromising the ability of future generations to meet their own needs."[34] But that fuzzy definition spawned a whole academic tradition aimed at trying to clarify what sustainable development actually means here and now. Thirty years on, there is still no agreed-upon clarity. So the word *sustainable* is now used to give any phrase a wishy-washy lift: sustainable growth, sustainable profits, sustainable competitive advantage. We also have sustained returns on investments, or even sustained high growth. The concept has been wholly appropriated by causes yearning for some embellishment.

More to the point, simply calling it *development* rather than *growth* for thirty years hasn't made the underlying growth go away. Growth is still with us, both in the human mind and in the economy, whether we like it or not.

This, then, is what I mean by the growth paradox: We can have neither growth nor degrowth, green growth nor qualitative growth. Nor can we ignore growth by labeling it something else. The UK economist Kate Raworth, author of *Doughnut Economics*, recommends that we become "agnostic" about GDP growth—neither believers nor disbelievers, but rather thinkers who can discern when growth is appropriate and when it is not.[35] The Dutch professor Jeroen van den Bergh recommends "agrowth" instead of anti- or pro-growth stances.[36]

It may be tempting to abandon the growth debate altogether. But the ideas of growth we citizens hold really matter to us, psychologically and emotionally, and thus they matter to the media and in politics. To parse out good growth from bad growth, we need to know what we think and feel, and why. Modern society is mainly built according to prevailing economic ideas as they determine what societies invest in. Thoughts in our brains eventually become physical external structures. If a large enough community of people imagine we want more things like cars, roads, and buildings—and the economic calculations show the benefits to be higher than the costs—then we will decide to create them. After constructing them, we start to inhabit these structures, and after a certain time of habituation, the current state of affairs becomes natural, with ever more cars and roads and buildings. Similarly, it feels safe and "normal" that GDP and incomes go "up"; otherwise people lose confidence that society is heading in the right direction. Images flow from the imagination into the world and back into our minds, solidifying our language and strengthening neural networks in our brain.[37] Our economic mind starts to expect more of the same. Rather than us consciously choosing our thought patterns, suddenly our habitual economic thoughts are subconsciously driving us. Now the economic tail wags the dog.

Once certain ideas about growth have settled in our brains, they work on us all by themselves until a new idea replaces them. Society forms the individuals who create society—forming a continuous loop.[38] Our social reality is constructed by such ideas going back and forth until they've settled in the minds of a critical mass of people. And then what we build changes. We forgo horse carriages or steam engines for cars and highways. After a while, we will cast off petrol engines for batteries, coal for solar power. How we organize work also changes: Organizations are no longer viewed as large, mechanistic, hierarchically controlled machines as they were in the Henry Ford era, but more like a network, a brain, or an organism.[39]

Now as we move deeper into the twenty-first century, new economic ideas may lead us to grow diverse cultures, rather than the massive mono-cultures of the industrial era, both in fields and in organizations. We can aspire to grow both crop *and* soil, not just max out on the crop yield while depleting what nourishes it. Sometime after such new ideas have settled in our minds, we may shift our habits around stuff, too—sharing more of it in a sharing economy rather than simply growing each individual's separate property. We may even grow a circular economy rather than a linear take-make-waste one—if that's what we imagine and start working to realize. Such is the power of imagination. It's time to harness its power to reimagine growth fundamentally.

UNDERSTANDING HEALTHY GROWTH

With either-or thinking like pro-growth *or* anti-growth, market *or* government, individualism *or* egalitarianism, the mind rolls down the slippery slope of polarization. We choose one side and feel we must reject the other.

"I do not believe that these critics [of growth] have struggled enough with the idea itself," writes cultural psychologist James Hillman.[40] If we just drop it, avoid it, or negate it, such dismissals of growth "do not satisfy the deep human wish that the term 'growth' symbolizes. To discard the idea only represses this archetypal desire and leaves it still encased in childish simplistics." Our alternative, then, lies not in denying growth but in reimagining and reworking both the imagery and conceptual analysis of growth. Reimagining it offers new openings, new generative metaphors, and new avenues for further economic growth and more constructive nuanced understandings of it. With that, our comprehension of growth itself may grow. Hopefully, we can grow richer by enriching our ideas of growth itself.

So, what to say the next time you enter a debate on growth? Ask two simple questions instead of being for or against it: *Growth in what?* That is, exactly which metrics are we speaking of? Money or tons? Value creation or wealth? Well-being or trust? And *what type of growth?* What image, property, or model of growth itself are we discussing? Gray or genuine green growth? Circular or linear material flows? Unfair or socially inclusive? Extractive or ecologically regenerative growth?

Many meanings and stories can be told of growth along several dimensions of value, and not one has a monopoly on the truth about growth. Yet in such debates I've found myself vying for healthy growth, by which I mean a more plural, balanced understanding of growth rather than the one-dimensional growth of sum total production output measured in market prices alone. And since growth is better understood as a multidimensional phenomenon, a complex network of relations, it cannot be reduced to a simple, juvenile polarity of up vs. down, smaller vs. bigger, higher vs. lower.

This type of healthy and enlivening growth is inspired by smarter designs that mimic nature and are powered by the sun. It builds on a broader, deeper understanding of money and capital, and it offers a regenerative model for balanced economic growth. This is the rapid growth of drawdown solutions that can start to reverse global warming.[41] It makes possible a simultaneous rapid degrowth of the most resource-demanding products in buildings, food, mobility, and energy. If carefully defined and progressively implemented, it can solve the dilemma of pro- vs. anti-growth.

Trees, for instance, don't seem to have a single version of growth. By the way trees grow, they contribute to better local livelihoods, ecosystems, and climate. They grow to their optimum height and then—rather than continuing to maximize height—grow into more unique, richer shapes and form elaborate rooted networks to other beings. Trees generate a lot of chemical emissions in the form of oxygen and litter, like fallen leaves and tree branches after a storm. Yet tree overpopulation is not a problem. Their emissions are a welcome addition to the atmosphere. Even the leaf pollution is productive in forests and manageable in populated areas. No, there's no tree overpopulation crisis in spite of there being around three trillion trees.[42] That's because trees are net beneficial to their surroundings, productively adjoined in diverse ecological networks. Humans, after all, are nature too. Why can't we imagine human societies more like trees or ants: billions of us, yes, and each of us net beneficial for our shared home, the oikos that our words for *eco*nomy and *eco*logy refer to?

The problem, then, isn't economic growth as such, measured in monetary terms by a company's value added or by the country's GDP. The problem is rather our oversimplified ideas about the content and type of growth that we've been creating. Our mental models of growth from the twentieth century are outdated and inadequate for the twenty-first. These previously successful ideas have made "us" humans (corporations and politicians in particular) oblivious to the extensive shadow sides of our actions. Instead, let's imagine that, like trees, we can revert to our own roots and find nourishment to reimagine growth more in the manner of forests: each factory a grove, each city a human forest. Then we can possibly find ways to realize growth in a healing way—grounded, regenerative, and enlivening.

Gray growth—the conventional economic growth model during the twentieth century in which environmental footprint worsens in step with the growth—was mainly an instrument in service of financial capital. Financial capital grew by "eating up" both social and natural capital. In contrast, healthy growth means the balanced accumulation of productive, social, and natural capital in which at least two objectively increase at sufficient rates of change without the third declining (more in chapters 5–7).

Healthy growth also extends to a wilder sense. Ecosystems run on both births and deaths, growth and decay. The decay in turn feeds the ecosystem as it evolves, like trees, into higher complexity and resilience. So, life-death-life cycles are growth cycles, too. The opposite of healthy growth, then, isn't no growth or degrowth or sick growth or decay. To me, it is a growth that is too one-sided. And one-sided models emerge when any one characteristic or metric (like GDP) is maximized and universalized at the expense of other types of value creation. Because this gives only partial maximization. We lose sight of what is an *optimum* state for the whole system. Then the one-sided bigger-is-better assumption dominates the entire skyline and horizon of growth (as well as voters' minds). It becomes fundamentalist.

THE ARCHETYPES OF GROWTH

In the tradition of archetypal psychology, the term *archetype* refers to the deepest patterns of psychic functioning. They tend to be more metaphors and images than things or neurons. We can't seem to touch one or point to one, even if we think of them as rooted in our limbic brain. You can imagine them as the roots of the soul governing our perspectives on ourselves and the world, going straight back to the early fairy tales, legends, myths, and cave paintings. Like literary genres, you can also find them in recurring typicalities in history, in the basic syndromes in psychiatry, or in the paradigmatic thought models in science: worldwide figures like kings, goddesses, superheroes, or monsters. You find them in films, rituals, relationships, and organizations. And they have an emotionally possessive effect in the psyche. Their bedazzlement of consciousness can make us blind to our own stances, writes James Hillman.[43] When falling in love, we experience the power of the Aphrodite archetype and become blind to the mortal shortcomings and shadow side of whomever we're enthused with. We can go crazy, turn wild, jealous, arrogant, or ideologically fundamentalist, all from the power of the underlying archetypal metaphor that rules our imaginative capacity and outlook.

By looking at what sparks disagreement over growth, we see some patterns emerge, and those patterns point toward four archetypes of growth: getting bigger, gaining mastery, producing abundance, and enhancing complexity. And they arise at the intersection of two factors: primary direction (up, down, around, or sideways), and primary properties (more or better).

Let's look at direction first. As I've explained already, to many people growth means *up*—a linear ascent over time, often in the short term. Think again of upward growth of the child. Or consider the ascent of the hero, or upward outlook of the optimist and investor. For others, growth is associated with a more recurring or cyclical movement; it comes, goes, and returns over time. This view holds that both increase and decay are part of the larger circular unfolding we call growth. Grass and flowers, for instance, grow "down" each autumn and winter, retreating into their roots, and then return "up" again each spring. As trees grow, they lose and regrow their leaves and strengthen their trunks. Salmon die after spawning. We humans, too, grow by letting go of old habits, outdated roles, and stifling patterns; selling off assets, or burying our parents. We grow by going back to our core being, seeking our true self.

Back to my hobbyhorse: "growth" is more suitable as a plural noun. Growth has different properties. Take size. Like a person grows in height, a city expands in area, or a company expands its production volume and number of locations. Let's call this property *more*. A second property isn't about becoming bigger in size, but rather about improving the extent of one's influence, excellence, control, and durability. This second domain is about being *better*. I can grow my influence in a company by moving up the hierarchy, becoming a domain expert or the boss. I can grow my mastery in speaking a foreign language, or fixing engines, or debugging software systems. I can also grow my power through more connections, wider networks, reaching new audiences, better communications. For this second property, growth is qualitatively better, not more of the same.

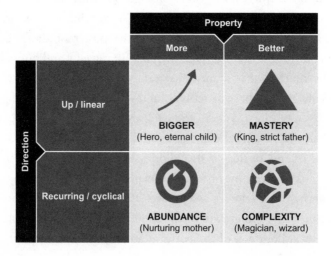

Figure 1.1
The four archetypes of growth.

These basic distinctions of direction and properties create a conceptual skeleton for four main types of growth: linear and more, linear and better, cyclical and more, cyclical and better. Let's flesh them out.

Linear and More (Bigger)

Combining up/linear with more recalls the heroic, simplistic, and childish ideas of growth. When we imagine growth this way, then more = bigger = good. It doesn't matter if you're talking about muscles, money, profits, or wealth; the size of cars, cakes, crowds, houses, planes, or farm herds; or horsepower torque, explosive force, or anything else. It's all good simply because it is growing bigger and/or higher. And the biggest also wins. The heroic archetype is about competing in size, strength, and speed, beating others and avoiding losing or falling down. In this view of growth, the world can be neatly divided in wins and losses. Losing is for losers. Heroes always win. Enemies are defeated. Bigly.

Linear and Better (Mastery)

If we combine up/linear with better, in the sense of better qualities like influence, we find images of growth that have to do with superiority and control. I'm now in better command of whatever it is that is important: people, soldiers, employees, funds, resources. The sizes may not be getting bigger, but my impact, career, and power grows. As I move up in the hierarchies of status or social power, I grow in reach and influence. Think of a king growing up and better at ruling his kingdom. Or a strict father[44] controlling his house, his family, his children, and his inferiors. The growth results in top-down control—over numbers, morals, or laws, for instance—and the power to define what is better in itself.

Cyclical and More (Abundance)

Unlike the controlling growth of the king or strict father, cyclic and more is about being fecund—highly fertile as in a lush garden, capable of producing abundant new growth. The cow gives birth and bountiful milk to her calf. Spawning fish lay millions of eggs that turn into millions of offspring. Most die, but enough are left to grow and prosper. Growth can happen by regeneration, recycling, rebirth. This archetype explains the emotional appeal of the circular economy as opposed to the linear growth model. Old clothes or carpets can be decomposed and born again as new garments, again and again. The European Union recently put out a directive called "closing the loop," a phrase that illustrates this archetype perfectly: Millions of tons of today's waste are to be remade into new materials, becoming useful and creating more value for each cycle, reborn "forever."

Cyclical and Better (Complexity)

The fourth type combines cyclical growth with better, as in the growth of complexity and of connections. Those who view growth this way do not envision rising up to the pinnacle of power, but rather growing through reaching out to wider networks, adding nodes to the web, weaving an ever finer mesh. The adult brain does not grow in weight; by making more elaborate neural nets, the brain's patterns get qualitatively richer. Also, as a system grows more complex and starts to connect its many parts in new ways, novel characteristics emerge. This growth of emergent properties can often take on unexpected and surprising characteristics. Not just the number of networked nodes, but communication itself can grow in the depth and multiplicity of its meaning. My grasp and understanding of a novel topic, for instance, can get deeper and transform through flashes of new insights that connect in better ways. The cyclical nature of how understanding grows has been described as following a circle of hermeneutic interpretation, as our attention moves back and forth between parts and the whole.

BRINGING THE ARCHETYPES TO LIFE

Using straightforward two-by-two characteristics to distinguish between four different types of growth models may be a useful starting point to break up rigid, monolithic notions of growth. But one problem with such an analytical tool is that it tends to hide the inner, psychological dynamics of how such notions work. Archetypal ideas are not just categories we can see on the page and fit well into four boxes. They are also entire ways of seeing and filtering the world. Ideas easily slip behind our eyes, into our psyches, and shape what and how we notice, the way we look at things, the way we argue. It's one thing when you have an idea. It's another altogether when the idea has you. There is a common saying that hints at this dynamic: "If you only have a hammer, every problem is a nail."

The ancient Greeks took note of the way these fundamental formative ideas can possess us humans, and they personified them into a cast of gods, so they could be more readily recognized and reckoned with. Each god reflected a timeless, archetypal quality of thought and emotion.

When I think of our first growth archetype, linear and more, two of these Greek personifications come to mind: Prometheus and Hercules. Prometheus was a titan who stole the fire from the other gods and gave it to humans so they could multiply and expand all over the world. He's a clever and busy engineer with little respect for the establishment, always able to design a technical fix for humans' problems, from metallurgy to masonry. His mindset can be expressed in a couple of defiant mottos: "Growth unlimited!" and "Grow or die!"[45] Hercules was primarily concerned with being the biggest and strongest. With his big club he could bang and bust any barrier on his road to growing his might and fame. His way up and forward was achieved by battling any challenger with his supernatural strength. Might made right. Competitions about who was the first, biggest, or strongest arose wherever he went. Their actions seem prescribed by something akin to a more = up = bigger mindset, in which defeat, degrowth, disease—or anything that starts with "de-" or "dis-"—is a threat. Progress means more of the same kind of growth, as much more as possible. If I thought in the age-old patterns of Hercules or Prometheus, I might think 3 percent GDP growth per year is not really enough; 4 percent per year is much bigger and better. And a 7 percent or 10 percent return on investment could still be quite disappointing, because the more = up = bigger idea has lured me into believing that it could have been ever more up.

The second vision of growth (linear and better) is more concerned about the long-term effects of rapid growth: booms and busts. Growth should be moderate, stable, and under control, toward better balance, clarity, and mastery. "All things in moderation," said the Greek god Apollo, and "Know thyself." He is the masculine ideal of hierarchical control and rationality personified. He is often seen alongside Zeus—the king of the gods, associated with stability, order, superiority, hierarchy, and sovereignty. Growth through a Zeus or Apollo lens would mean that power is increasingly structured and centralized in a few hands, well managed, and stable over the years. The Zeus archetype supports a positive view of the establishment, with rational laws set by a strict but just king. Perhaps this king of the gods would argue that a moderate, balanced, and controlled 1.5 to 2 percent annual GDP growth is preferable to wild fluctuations and uncontrollable busts and depressions that need intervention and top-down corrections. The exalted role of the governor of the central bank, as the king of the money system, may serve to personify this idea.

The third vision of growth, cyclical and more, reflects fertile abundance, endless coming and going, the birth and nourishment of all things growing. This is a more rooted perspective on growth, closer to natural cycles and ecosystems. It is local and seasonal. These ideas are often associated with the green goddess Demeter and her grandmother Gaia, who nourished and fed all living things, from seedlings to siblings. We could imagine them as the green photosynthesis personified. From this perspective, we're all grass. We're fed by their grain, fruits, and leaves, drawing nourishment from soils and roots. Growth, in this ideal, is more bottom-up than top-down. But the abundance it produces still has limits and is exhaustible. When someone sees the world through this lens, they see Earth's fertility threatened by one-sided bigger = more policies and huge, centralized, monolithic megastates or megacorporations. The environmentalists' vision of growth is often closer to the ideals expressed by Demeter and Gaia than by Prometheus, Hercules, Apollo, or Zeus.

Finally, the cyclical and better idea of growth tends to see growth in light of ever smarter configurations. Growth here no longer refers to the "gross" up = bigger, but to an endless increase in complexity and dynamism. This is a growth in creativity and acuity, not in brute size, control, or fertility. The Greeks probably would have associated this never-ending movement from simple to complex, from mono- to multidimensional, from constrained to transgression and transformation, with Hermes, the herald, the thief, the trickster. Even in his epitaphs he is multiple, complex, a shape-shifter. Hard to pin down. He is depicted as unpredictable but funny and smart. He connects what was previously unconnected. He grows networks, endless virtual realities. We can imagine him running the digital domains, the unpredictable AIs.[46] He also guides the souls over the borders, shifting seamlessly from one domain to another. When we see growth through this lens, we see a progression from simple to complex, from isolated to connected, from brute force to smart dealing skills, from confined to transformative. One of his characteristics is the infinity symbol, ∞—which when related to growth means neither up nor down but never-ending cyclical transformations, forever branching out.

Pairing our four archetypes with Greek gods—Hercules (bigger, stronger), Zeus (mastery, power), Demeter (abundance, regeneration), and Hermes (smarter, complex)—might seem flip, but the point is that there are age-old metaphors at work deep in the human mind when it considers complex phenomena. Those metaphors become even more programmed into our thinking when reinforced by the Western Greco-Roman languages.[47] They refer to characteristics of human thought and emotion that don't change that much from century to century, because the basic structure of our brain remains the same. Knowing about these patterns can help us recognize different qualities in the images of growth that we subconsciously hold, images that often create conflicts and that may even lead us astray.

Sometimes, both individuals and whole societies can get possessed by an archetypal idea. Wrote Jung, "If the affective temperature rises above [a certain] level, the possibility of reason's having any effect ceases and its place is taken by slogans and chimerical wish-fantasies. That is to say, a sort of collective possession results which rapidly develops into a psychic epidemic."[48] In this mental state our sense of cocksureness can get inflated beyond all reasonable bounds. Some critics say that Western consciousness has been addicted to, or emotionally possessed by, the endless growth story. It has become a full-scale dogma, an unquestionable paradigm.[49]

Therefore, it's important to recognize how ideas and metaphors become social constructions that then shape our shared knowledge and the physical world.[50] One key to succeed in changing the direction of growth may be found in becoming aware of what underlies our ideas about it. That means self-reflection about the images and narratives that guide us when we speak of it. And paying attention to which direction and properties we expect growth to take. Together they open up to a new psychology of economic growth: not just the numbers and the equations, but the metaphors and emotions too.

But if one is in the grip of growth addiction, resistance to change gets tough. When companies producing cars, coal, oil, cement, or steel can see no future other than the linear growth model of more the same, when investors want nothing else than to grow their own assets way beyond any level they need to live, then we are in the growth addiction zone. New directions are hard to see, and satisfaction from "enough" is nowhere to be found. Our normal way of growth becomes the only way forward, come hell or high water. Then the consequences of that addiction begin to grow, too, becoming visible as glaciers, forests, and pollinators disappear and soils turn toxic from chemical fertilizers, heavy metals, and excess tilling while the oceans slowly rise and groundwater aquifers deplete.

Luckily, there are better forms of growth than the one that is currently deranging forests, rivers, oceans, and the climate. There are also better designs for capitalism than the one that is currently ripping apart societies, escalating inequalities, and undermining democracies in pursuit of its one-sided goal. Rather than negating capitalism and trying to stop the growth train in its tracks, we can rethink growth so that capitalism starts to serve society and nature rather than dominate it. We can still set it on a new track, a new direction. We can redesign and redirect the momentum toward a healthier growth.

But the process necessarily starts in our psyche, not just by reimagining the images we hold of economic growth but also by reimagining capitalism. As the author and futurist Alex Steffen has observed: "It's literally true that we can't build what we can't imagine. . . . The fact that we haven't compellingly imagined a thriving, dynamic, sustainable world is a major reason we don't already live in one."[51]

FROM NEGATING TO REDESIGNING CAPITALIST SYSTEMS

Many left-leaning environmentalists say that capitalism is the enemy, corporations are to blame, and the corporate capitalism they've spawned is utterly broken and killing both people and planet. They call for system change—not climate change. Naomi Klein pits capitalism against the climate.[52] Yes, the current version of capitalism may be wreaking havoc, but it's not that capitalism is broken. Rather, it is doing exactly what it was set up to do: deliver return on investment to capital owners by serving human wants through the market. The problem is that capitalism is working one-sidedly and *too effectively* in this narrow sense. It is today, in the first decades of the 2000s, a runaway juggernaut that we now need to learn how to contain and handle safely. If we want change, we need to redesign the framework around capitalism and growth, not negate, bash it, or kill it, denying the human psyche its subconscious yearning for growth. Because neither capitalism nor growth dynamics are going away anytime soon.

We should acknowledge that capitalism and markets are among humanity's greatest ideas and achievements. But we must be clear about what they are—and what they are not. At its core, capitalism is a *social engine*, a blind mechanistic system: It accelerates investments to increase profits and further asset growth for evermore investments. Its root, *capital*, refers to both tangible, manmade things like steam engines, railroads, cows, and computers and to the funds needed to build and purchase those things. As we know, capital begets more capital. And it lets the strongest players, those owners with excess money, put their capital into action to make more stuff: goods and services. Those who put the most money in get the greatest returns back out (on average). Fundamental ideas such as property, assets, interest rate, and return on investment make sure this happens predictably, year by year.

That's why the haves can make more money than the have-nots. And it is an exponential affair: 10 percent interest on $100 million yields $10 million per year in return. But 10 percent on $100 yields only $10 over the same period. After seven years with compound interest each investment will double. Our have now owns $200 million; our have-not only $200.

Escalating inequality is inescapable with that kind of math. Today, 84 percent of all stock value is owned by the wealthiest top 10 percent of the population.[53] This sad (for most people) consequence fuels increasing inequality in fundamental ways. You could label it unfair greed, but in truth it is mainly mathematics embedded within mechanical, cold power structures (like banks, software, shareholder registers, instruments of debt and money, and accounting laws) that maintain exclusive ownership to capital and its returns. If you own a large, broad investment portfolio that is well setup, it keeps accumulating as you do nothing. Until you actively withdraw funds. Or give it away. Such is the DNA of capitalism. It enriches mainly those few who already have more capital than they need in their daily lives.

Conversely, the more you need money to stay alive or even start your own business, the more you have to pay for it. Thus, capitalism predictably transfers money from the many poorer (the net debtors) to the few rich (the net creditors). Capitalism and credit works because the system, by design, ensures that all the money borrowed by have-nots is returned, with interest, to those who already have accumulated the most. Already in the 1850s Karl Marx saw that capital is "money which begets money." Thus, complaining that capitalism creates inequality is a little like pointing out that water creates wetness. And complaining that capitalism is using too many resources to satisfy its "greed" is like complaining that kids take too much ice cream at an all-you-can-eat buffet. Of course they do.

Capitalism has self-reinforcing feedback loops, running faster and faster. It does not know when to stop, and it has no concept of decency, or an optimal level of money or wealth. Capitalism—as a system of production and consumption—does not come equipped with brakes. Wild swings in torque, yes—sometimes sputtering, reversing, at other times violently revving. But brakes? No. It is always bent on maxing out growth. A sense of justice? Gratitude? No again. Dignity? Ha ha. Only if it helps the next deal. In business you never get as deserved; you get as negotiated. Capitalism does not know how to share the amazing amounts it amasses to the lucky, successful few. On its own, capitalism is simply an impersonal, mechanical, and exponential engine. A spinning wheel, doing its thing.

It is up to us to shape and contain it, give it direction, supply it with brakes and balance, institutions and norms. This can rein in runaway acceleration and return a fair share of the benefits to the commons for the greater public good. A better-designed framework can keep the dynamic from destroying itself through its own manic excesses in wealth creation and concentration. For the incredible vitality and power of capitalism is also its own worst enemy.

The real argument is not about growth or degrowth. It's not about capitalism or the climate. It's not about money or your soul.[54] It's not about finding an alternative to capitalism. The key issue is how to shape a highly diverse set of capitalist systems into smarter and more balanced versions that actually support *and* serve different societies and cultures while also restoring natural areas. It is about redesigning what we have so it doesn't rip our earthly home to pieces in a gigantic, blind binge. Rather, let's reframe growth from being about the *gross* size to growing healthy smart designs. Let's grow out of the linear "more = bigger" idea of growth and introduce more cyclical, complex ideas of what growth is. With increasing wealth it's time to shift from Hercules to Hermes, to use ancient Greek metaphors.

Actually, as with growth, a singular thing called capitalism doesn't really exist. There is no pure capitalist system anywhere in the messy empirical world. Instead, there are many culturally and socially distinct capitalisms. Each version is embedded within a society that also always is shaping the core engine one way or the other. There are already many cultural varieties of capitalism in our world: The American type of capitalism is very different from that of China, the United Kingdom, Germany, Brazil, the Nordic countries. And inside these nations, subnational capitalist systems work very differently, each with its own unique history and evolution. The Silicon Valley version, for instance, differs from the Appalachian version. They will never settle into one ideal model or mold, as in the neoclassical fantasy. And there is no single fix for all the varieties. Each capitalist engine is contained and shaped by a unique set of laws, norms, and other social institutions.[55]

Markets can be incredibly vital creations. They make for freedom, fun, and dynamism, as well as ruthless creative destruction, win-or-lose dramas, abuse of power, or corruption. They are great as engines to generate momentum, wealth, health, interaction, and surprises. But they are terrible as masters of our mind. The capitalist market system of consumers and producers repeatedly ignores what direction is socially ethical or what scale is ecologically justifiable. There are markets for slaves, sex, child labor, drugs, assault guns, and rhino horns. Capitalism by itself can never know what is enough and sufficient. Only democracy, ethics, science, and governance can balance its inherent expansion. Only forces external to the core engine can shape market growth in new directions and give it some sense of fairness.

In Greek mythology, Hermes was the god of commerce and markets. (In Latin his name was Mercury, from which came *merchants* and even the word *markets* itself.) And the god of science was Apollo, the clear-sighted, lauded, rational one. Surprisingly, they were half-brothers. Hermes, god of commerce and markets, brother of Apollo, god of science, clashed and negotiated in a forum on Olympus as heard by fellow gods and governed by their father and king of the gods, Zeus. Maybe we could rethink growth in markets more through the lenses of Hermes than of Prometheus or Apollo? That would mean understanding markets as conversations, not just mechanisms of growth or perfect competition.[56] Markets are—in the light of their archetypal origin Hermes—ways of valuing and negotiating for mutually beneficial exchanges. In markets, Hermes-type thinking fosters communication, relationship building, networks and innovation.[57]

These concepts go way back in the Western culture and subliminally guide modern ways of thinking. But we—that is, anyone participating in this modern economy—could do well to remember, reframe, and reorient ourselves after the series of booms and busts since 2000. As a society we have still not learned how to master growth. After the Great Recession of 2008, all efforts went to chasing some mythical "normal" growth.[58] The question now, after the 2020 crisis is how we heal growth at a deeper level: Instead of getting back to the old normal, just growing more, faster, bigger again, it's time to critically appraise the track we're on. It's time to reimagine and retell the story of growth itself.

2 SOWING THE SEEDS OF TOMORROW'S ECONOMY: CREATING HEALTHY GROWTH

What if it were possible to produce cars, buses, foods, and buildings that actually cleaned the air? What if products were made from materials that can be used over and over again, while doing no harm to the ecosystem nor to the workers who make them? Imagine shoes and shirts you can return or compost once worn out, biodegrading into better soils. What if all the world's vegetables, fruits, cereals, grains, legumes, and nuts were grown in ways that create new, healthy soil rather than poisoning and depleting the soil? What if the typical methods of meat production didn't pollute waterways, divert a massive amount of feedstock, release dangerous levels of greenhouse gases, or cause animal suffering?

If the above opportunities were enacted, more human consumption could have a net-positive effect on nature. Our environmental footprints would in sum first become less destructive, and then become net beneficial to our surroundings. Growth—for a company or for an economy—would first kill off wastefulness, and then help repair our world. Economic growth starts to regenerate soils, forests and rivers.[1] Are such ideas feasible, and if so, might they even be more profitable?

It seems increasingly likely to me. But before assessing those ideas, we need to identify and transcend the fuzzy area between greenwashing and transformation. Luckily, a swelling swarm of eager eco- and social entrepreneurs are on it. Let's visit a few of them.

When UK activist and author Tristram Stuart founded Toast Ale in 2015, he had already spent many years building awareness about the impacts of conventional food production. In the course of kicking off what became a global movement against food waste, he began to understand the enormous business potential in recovering even a fraction of our colossal food waste and converting it back into valuable products. But it wasn't until Sébastien Morvan of the Brussels Beer Project showed Stuart how to make beer out of leftover bread that he decided to dive into a business venture.

Stuart had seen vast industrial quantities of waste bread all over the world. He also knew that the virgin grain used in traditional beer making was a mostly conventional affair: grown in ways that have made our food system a voracious consumer of freshwater and an enormous contributor to deforestation, greenhouse gas emissions, soil erosion, and species extinction. Here was an opportunity for impact. He jumped into the fray of startups.

Today, Toast Ale's beer is made in seven countries, using waste bread rather than virgin grain to drive the fermentation process. The company, a certified *B Corp*, has made its recipe available to anyone who wants it, and it gives 100 percent of its profits to a charity fighting food waste.

Says Stuart, "All of my work has been to upcycle what in narrative terms is a catastrophe, an ecological and social disaster—namely that we waste a third of the world's food—into something that can inspire us."[2] Toast Ale turned out to be a very concrete, business-based way to tell that upcycling story. But, like many leaders firmly engaged in the circular economy, Stuart asked himself hard questions about producing a product: How much growth is too much? At what point could we end up doing more harm than good? Are we producing stuff no one really needs?

"Beer is a luxury product, and for the sake of the environment it would be better if everyone drank less of everything," he acknowledges, meaning everything that comes from a bottle or can. "We need to replace normal beer consumption with waste-bread beer consumption. That will reduce the land and other inputs that go into producing the grain that goes into beer." In other words, Stuart's vision of growth is to create more value by converting slices of the beer-demand pie rather than expanding total consumption. "If you could get your hands on all the waste bread produced in the UK, then you could produce nearly all the beer consumed in the UK using our recipe. This is a huge industrial opportunity." Stuart had come across a way to outcompete conventional beer with a low-footprint challenger.

So, is Toast Ale simply doing less bad with its share of the beer market than its virgin-grain-using competitors, or is it doing actual good? "There is a lot of misguided, counterproductive nonsense being peddled out there in the name of saving planet Earth," says Stuart. Frequently, he says, the very companies touted as sustainable or socially responsible are "sociopathic monsters. Not because of anyone's fault, but because they exist in a legal structure that obliges them to produce maximum return to their shareholders." That demand for short-term profits drives companies to what Stuart terms "ecocidal behavior."

To Stuart, drawing the line between greenwashing and genuine green transformation demands that we look at more than just a company's physical product: "There's another product that all these companies are producing that has not received anything like the kind of scrutiny that the ostensible products of these companies have received. And that product is money." A company might strive for a circular material flow, he notes, but still have a linear money flow. "Let's say you set up a company and sell it to Unilever for $100 million, and then you walk away with that cash and invest it in an oil company or flights to Barbados. Like the vast majority of money in the economy, it is doing harm to the environment. Only small pockets are doing good. So if you just have money that flows back into the main economy, how much of the good that your company has done is being offset by the bad that your money is going to do farther down the line?"

The fact that Toast Ale commits 100 percent of its profits—unique among corporations—resolves only some of this concern. Investors could still, through the sale of the company or the sale of their shares, cycle their earnings into "gray" versus "green" ventures. To avoid this, Toast Ale had all of its investors sign an Equity for Good pledge, stating that any money they accrue through capital gain on the sale of the company or their shares has to be reinvested in organizations, whether for profit or nonprofit, whose core business is to save the planet.

"We all buy into companies that we know are dumping their waste and their negative impacts onto the environment and onto society," Stuart reminds us. "They are not paying their external costs. We, the citizens, are. We pay twice—we pay for the things we are buying, then again when we pay for the environmental or social damage they do. That is a supremely inefficient way of buying into an economy. How about instead we design and invest in businesses that do the exact opposite: They produce social and environmental dividends with the money we are giving them. It's a much more sensible way of running an economy."

LEAPFROGGING FOSSIL FUELS WHILE ELECTRIFYING THE WORLD'S POOR

Those kinds of social and environmental dividends drive yet another enterprise that is attempting to do business differently: Solar Sister, a network of women who are lifting themselves from poverty while advancing their communities in sub-Saharan Africa.

In rural eastern Nigeria, Eucharia Idoko is one of many who long lived without access to electricity. The power grid in her region went down in 2004. For years, nights were dark for this mother of seven and grandmother of one. And in her community, the darkness invited danger—roving gangs that would approach houses, kidnap residents, and return them for ransom.

Yet by the time Idoko's house was targeted, she had become a Solar Sister, trained to sell low-cost, solar-powered lights, fans, and cell phone chargers to other women in her area. When a gang showed up, Idoko's solar lights went on, and the possibility of the kidnappers' faces being illuminated kept them at bay long enough for her to call for help.

To Idoko—one of about four thousand Solar Sister entrepreneurs who have brought clean energy to more than two million people in Nigeria, Tanzania, and Uganda—light brings more than the ability for children to study at night, or for health workers to navigate clinics in what would otherwise be darkness. It also brings security.

She and the other Solar Sisters live in a part of the world where 600 million people lack access to electricity. Because more than 700 million there cook over open fires fueled with wood or coal—causing health issues as well as environmental woes—Solar Sister entrepreneurs also sell clean cook stoves, which use far less firewood, far more efficiently, for each meal.[3]

Solar Sister cofounder Neha Misra says the grassroots initiative, funded by both philanthropic support and product sales, fills a critical need—and a certain kind of economic growth is required to propel it. "We need big solutions for cities and we need solutions for some of the biggest greenhouse gas polluters in the world, but let's not forget that the most underserved, most vulnerable communities in the world have been waiting for many, many generations for electricity that has not come yet. So how do we change that? The largest source of renewable energy in the world is women power, and it's one of the least tapped sources."[4]

Her organization scouts local women leaders; provides them with a year of training in marketing, record keeping, and other business skills; equips them with the tools and training they need to repair what they sell; and offers them other ongoing support.

The goal is bigger than spreading renewable energy, reducing illness from woodsmoke, or the other more tangible benefits of their work. "Women are fundamentally critical partners and changemakers to build a more feminine economy," explains Misra, who hopes we can do away with what she calls "the false dichotomies of the past"—like economic growth *or* environmental well-being. "What we are trying to do is change from the board room to the village to build a more inclusive economy of the future."

By 2024, Solar Sister plans to increase their network of entrepreneurs to 10,000 women in sub-Saharan Africa. If that happens, estimates the organization, it will have brought clean energy into more than ten million people's lives.

For this kind of scaling, they need more funds. One study from consultancy McKinsey estimated the potential addition from increased women's participation in the global economy to be an astonishing 12–28 trillion dollars per year.[5] And for socially and environmentally driven enterprises like Solar Sister, those funds may be arriving through innovative channels too.

I first met Sam Manabieri, founder of Trine, when I was giving a TEDx talk in Gothenburg on our "sun-rich future."[6] In the talk, I had mentioned his company as one of the healthy growth companies enabling such players as Solar Sister. When Manabieri came up to me after my talk, exuding enthusiasm and optimism, I learned a bit more about this bright, young Swedish engineer who turned himself into a finance expert with a mission. Trine offers people with savings an opportunity to let their money work directly for people like Eucharia and the remaining hundreds of millions still lacking electricity for lights, charging, and security. They leverage crowdfunding from a range of sources, from citizens with extra funds to large institutional investors. He and his team have figured out how to crack the hard nut of crowdfunding better than most, and their company is growing at breakneck speed.

Some months later, when I was serving as member of Norwegian parliament, Manabieri came to Oslo. I had invited him to the parliament to discuss opportunities for healthy solar growth based on crowdfunding combined with new funds from larger players, such as the Norwegian sovereign wealth fund. I decided I wanted to put a substantial part of my private savings into this healthy growth financing model while also working politically to shift the billions in the Norwegian sovereign fund away from fossil investments and into renewables. Trying to put my money where my mouth is, I want to support the acceleration of these financial business models for healthy growth. Trine's business model offers not just better interest than my bank but also more secure dividends than owning stock shares. This is true impact, healthy growth investing.

A SUSTAINABLE PIONEER RAISES THE BAR

As exciting as these startup collaborations between business and philanthropy are, healthy growth can happen in more traditional industries as well. Just over twenty years ago, Ray Anderson, then CEO of Interface, one of the world's largest carpet manufacturers, became an unparalleled champion for sustainability after reading Paul Hawken's *The Ecology of Commerce* and having what he called his "spear in the chest" moment. Recalled Anderson, "I . . . was dumbfounded by how much I did not know about the environment and about the impacts of the industrial system on the environment—the industrial system of which I and my 'successful' company were an integral part. A new definition of success burst into my consciousness, and the latent sense of legacy asserted itself. I got it. I was a plunderer of the Earth, and that is not the legacy one wants to leave behind. I wept."[7]

Anderson also leaped into action, assembling a dream team of environmentalists, renewable-energy experts, biomimicry pioneers, and others to help him achieve a company overhaul that would move Interface toward an ambitious goal: to have zero impact on the environment by 2020. When he died in 2011, his company had already made substantial progress toward that goal. And while the company won't attain all of its mission zero targets by 2020, they have come farther along the sustainability journey than any other large industrial company out there.

They've set their goals a bit higher, too, heading in the direction of healthy growth by asking themselves a simple question on the climate front: Why aim for just zero emissions when we could aim for a climate-positive impact instead?

According to Interface's chief innovation officer John Bradford, the company wanted to "take carbon dioxide and turn it into a mineral-like form. There are a lot of things in the world that do that very well, and one is called a tree. It breathes carbon dioxide and turns it into a sugar. Our goal is to create processes to take carbon dioxide out of the air and put it into our product in the form of a mineral or plastic that will stay there for over 50 years."[8]

John Lanier, Ray Anderson's grandson and head of the foundation that bears his name and spreads his vision for industrial overhaul, reports, "Less than a year after Climate Take Back was launched, the company had developed a prototype carpet tile with a negative carbon footprint. Specifically, the backing of that carpet tile sequestered more carbon than was emitted in the full lifecycle of the product. In 2018, Interface announced that all of its products going forward will be carbon neutral." Some products depend on carbon offsets for that neutral status, he added, but the company estimates that those offsets won't be needed in the near future. Notes Lanier, "Every sale of Interface carpet will mean that a little more carbon dioxide leaves the atmosphere."[9]

Interface—which has been a leader in incorporating biomimicry for sustainable advances—has also embarked on a new initiative called Factory as a Forest with the consultancy Biomimicry 3.8. "The idea is simple," explains Lanier. "Interface wants its factories to generate all of the ecosystem services that the land would have provided if the factory had never been built, including carbon sequestration."[10]

Such grand plans often have difficult passages through board rooms, and Interface's experience has been no different—in the past, or now. Despite some skepticism, the company quickly saved hundreds of millions of dollars instituting its first sustainability measures years ago—many aimed at waste reduction. But efforts to erase footprints altogether or transform materials into carbon sinks come—at first—with hefty price tags. Breaking away from well-proven tracks of the mainstream often comes with substantial financial risks. Despite some raised eyebrows, the Interface board chose the bold path forward. If the future mimics the past, the company will reap financial rewards even as the rest of the world catches up. Since 2000 and up to the time of writing, Interface has outperformed the main stock market indexes.

TRANSFORMING BUILDINGS INTO POWERHOUSES

It's not just old manufacturers that need to grow and prosper in new ways. We also need a growth spurt to make our unhealthy buildings—responsible for nearly 40 percent of climate emissions—into healthy ones.[11]

In the mid-1980s, when the oil company Saga was drilling briskly in the North Sea, it found it needed new headquarters. So it purchased land outside of Oslo and erected a building with that period's fashionable materials: dark glass, metals, concrete. Since the designers weren't optimizing for energy and daylight, they shoved in lots of incandescent lights, heavy-duty ventilation, and heating and cooling systems to regulate the indoor environment by sheer brute force. By 2010, the oil company was long gone. But the building remained standing, though it was no longer suited for office work.

Six years later, it had been transformed. Today, the design and lighting are warm and inspiring, and 95 percent of all materials are either recycled or recyclable at the end of their lifetimes. In the headquarters of Asplan Viak, overlooking the Oslo fjord, porous white wafer plates hang down from the exposed concrete ceiling, spreading light in a wavelike pattern and giving an airy feeling to the room. The plates are made of 100 percent recycled PET bottles, noted Øyvind Mork, former CEO of the company. "We employ the concrete in the walls and ceilings as heat buffers both in summer and winter, so that's why we don't want to cover the concrete. And by having these cloudlike wafers in front, we get great aesthetics and acoustics, and terrific heat management over day and night."

The building's transformation took place after some visionary folks got together at Norway's annual zero-emissions conference back in 2009 and decided to form a partnership that included a construction company, the real estate owner, an environmental organization, an aluminum profile company, and Asplan Viak, an architect and engineering consultancy that needed new headquarters. They wanted to transform that 1980s office building into a powerhouse for the 2000s—a building that generates more energy than it needs for its own operations and more energy than is embedded in all its materials—meaning more energy than the building uses, and more energy than it took to extract, manufacture, and transport everything it is constructed from.

Those plans birthed a world-first total refurbishment of an old, worn-down office building into a "plus-house" and an inspirational workplace. The project was nicknamed "Powerhouse."[12] Mork, at the time also chairman of the Powerhouse consortium, said that Asplan Viak had been renting the building next door when the owner called him, right after the group had formed at the conference. "The question was if we would want to technically draw and execute the project, and then move in as the first tenants. We didn't ponder long."

Before moving into the Powerhouse, recalls Mork, "when my colleagues and I went to work, we'd be consuming energy. Now, when I go home in the afternoon, while planning better buildings for others, we have generated more energy than we consumed during the day."

What about cost? "The first Powerhouse we built ended up being 10 to 15 percent more expensive than a rehabilitation according to Norwegian existing standards," Mork explained. Some of that extra cost was born of the initial learning curve, and from the high aesthetic ambitions of the architects, the world-renowned Snøhetta, who wanted rounded indoor walls and an open, oval staircase as the central air shaft of the building. "Now we're doing the second Powerhouse here, in the neighbor building," said Mork. "We see this new way of thinking and rehabilitating buildings is approaching zero extra costs as we learn from experience. Look at it this way: You get much better buildings, at no extra costs, which are cheaper to operate and make more energy than they need over their lifetimes. . . . What's the argument for building the old way?"

Mork expects the Powerhouse concept to spread rapidly. "We are aware, of course, that real estate is a very conservative industry. And some of its most backward players just made it to run government. Yet the business logic of this new innovation will soon make it unstoppable. It's a matter of spreading the word, examples, and education. Seeing is believing!"

LEANING INTO RISK

The desire to build a truly regenerative economy is spreading. Many are now hard at work changing buildings, transportation, industry, food, and waste handling—the five main end-user sectors of the economy that together cause—according to the Intergovernmental Panel on Climate Change, IPCC—nearly 100 percent of all greenhouse gas emissions.[13] Scientists, entrepreneurs, corporate rethinkers, makers, farmers, architects, builders, and a host of others are developing ways to deliver what people want, materially, in ways that either do no harm or even provide net-positive results.

Like the Climate Take Back team at Interface, Solar Sister, Trine, or the Powerhouse designers at Asplan Viak, they believe that we can transform our economies by rolling out new net-positive products and services—and they are creating new business practices and models to do so. To become transformative, though, they must drive massive change. And to drive massive change they must be profitable enough to scale up quickly to outcompete the older "gray" solutions. Trusting that this kind of net-positive enterprise is possible relies on believing in the quality of the early innovators and early adopters. If transformation happens, it will be because entrepreneurs and intrapreneurs actually get their acts together, forging ahead with ideas they think can become part of the larger puzzle, picking up speed and momentum as they go. But it will also take many brave starts, many redirections, and new approaches to research and financing.

What I'm speaking of here is not just words or academic phrases, but lessons I learned early on when working for more than six years to bring one healthy growth idea to life. In 2008, I cofounded a company called GasPlas. The idea was to use microwave-induced plasma to crack fossil natural gas into hydrogen gas and solid carbon in a powder form. The hydrogen could be used to run vehicles that emit only water from the tailpipe—achieving close to zero emissions for trucks, buses, taxis, and cars. The carbon could be used in construction materials or batteries, or be applied as a soil enhancer.

Even better results can be had by using biogas instead of natural gas. Conventional hydrogen production from fossil gas generates a lot of CO_2 at the refinery (a problem) or uses copious amounts of electrical power (another problem) for the electrolysis of water. Biogas, on the other hand, is generated from food, agricultural, and other organic wastes, along with sewage. To make it, the organic matter is digested by anaerobic bacteria in a closed chamber. From this chamber come large amounts of methane (CH_4) that can replace fossil gas. And here's the extra point: If we cracked this biogas using microwave plasma, we'd get hydrogen out with no CO_2, only solid carbon at the fuel station. You could then fuel a vehicle with that green hydrogen, emit only water, and hence clean the air for CO_2 as you drive.

If the carbon powder were put into durable materials or soil, then the whole process—and the driving of trucks, buses, or cars—would become net-positive for the climate, leaving less CO_2 in the air with each mile driven. Because biogas is generated from organic waste, it could become cheaper than fossil gas. And if the carbon materials could be sold at a good price, too, the overall economics point to strong profitability.

So, driving with hydrogen made from biogas would be better for our wallets, better for the air, and better for waste handlers. It would also create more jobs with declining emissions and higher profits. That's what we mean by healthy growth. Is it too good to be true? And if this idea is so smart, why isn't everyone doing it already? The answer has a lot to do with the process of healing our old growth habits, confronting the challenges and barriers to green innovation, and considering how we analyze overall life cycles. It also has to do with surprises—and why the future is inherently unknowable.

My own story is a case in point.

VEHICLES THAT CLEAN THE AIR?

The main idea of GasPlas was simple: take the carbon out of hydrocarbon gases, leaving hydrogen for cars to run on, water as the only emission, and carbon materials that can be put to long-term uses rather than burnt and pollute the air. I wrote a research funding proposal, outlining the previous state of the art of producing microwaved-induced plasmas and why our new patent-pending reactor could improve it. The basic plasma process was well-known to lab science, but no one had yet solved the practical problems of handling the carbon it produced when brought to scale—in other words, when producing large volumes. Our idea was to solve that problem through a new reactor design. We started building in our garage lab in Norwich, UK. The early tests from our proof-of-concept reactor had indicated that we could actually do it. Half a year later, we won public-project grants sizable enough—when matched with private funding—to start building the pilot reactor. Everyone on the team was thrilled.

The first three years of the GasPlas venture were a whirlwind of opportunities, vision, teamwork, negotiations, setbacks, and successes. Some days, when we were achieving breakthroughs, the road ahead seemed rosy, and we could envision ourselves as captains of industry, disrupting the entire hydrogen and transport industries. Other days, when we were discovering errors in our designs and calculations, we despaired that we'd launched into a foolish, doomed endeavor.

A few years in, we were headquartered in Oslo, with labs in the UK, on the University of East Anglia campus. Our results were good enough to secure a few more investments and major R&D grants: one to make a hydrogen-production prototype, and another to optimize the carbon by-product. Brimming with hope, we set about building the necessary advanced lab infrastructure and designing the prototype plasma reactor. I made a contract with a landfill operator that could supply ample biogas to our prototype, as well as a biogas sludge digester and a compost process that could be improved with added carbon powder. In this way we could demonstrate the whole value chain: from waste to biogas to hydrogen and carbon, with the carbon permanently stored in improved soil. We even came up with a new name for the process: Not the conventional carbon capture and storage (CCS), but carbon capture and use (CCU). I enthusiastically registered the website carboncaptureuse.com in expectations of a wildly successful future.

With this exciting prospect, we attracted a lot of industry interest; there were meetings, calls, journalists, BBC World reports. Some of the largest corporations in the world wanted to come visit. Several of them signed nondisclosure agreements, project proposals, and research agreements.

But after hiring a lot of people, progress on engineering the plasma reactor prototype was not satisfactory. We missed milestones. Worse, the milestones weren't clearly defined. And there were unexpected setbacks due to health, safety, and environmental requirements that our lab couldn't easily fulfill. You can't let your workers mess around with a combination of explosive gases, high voltages, intense microwave radiation, unknown chemical processes, and high-pressure equipment! The number of detailed regulations—the checklists, certifications, inspections, and approvals—boggled the minds of unprepared, eager entrepreneurs. Disagreements over designs, software development, safety standards, zoning, working conditions, and project management philosophies erupted, some with considerable animosity, requiring all my skills as a communicator and psychologist to calm heated minds and redirect the team toward achieving common goals.

This caused a lot of lost sleep. I was in the midst of it all, no longer the calm external consultant or distant academic. After several painful reorganizations, we achieved our objective: The prototype reactor was ready to test. With lots of tweaking, the plasma reactor successfully cracked the gas to hydrogen and carbon at scale and continuously. Now our main problem was that we needed a good off-the-shelf carbon filter to ensure that the hydrogen we made was squeaky clean, so that it would not contaminate the sensitive fuel cells of hydrogen cars.

We ordered a top-level, expensive carbon filter from a leading industrial supplier to get the job done. The supplier promised the job would be run of the mill for them, but then months passed. No filter. The supplier seemed unable to provide what we needed to commence our testing. They were suddenly very busy with larger customers. Finally, nine months delayed, they delivered the final parts for the filter, and we started testing. It seemed to work fine. Again, elation.

But then—in the final months of our project funding—the filter started to clog. More testing and tweaking ensued. We tried different filters, eventually arriving at a very expensive option that worked only for a few hours, then stopped. Which means it was useless for industrial purposes. We had tried all the easy solutions. The next step would have been to launch a new, separate R&D project to solve the carbon filter problem. Which would have meant more funding. Without having secured success on our first project, that would be difficult. At this point, my hair had become considerably grayer. Then the CEO resigned, and the project leader went to Thailand for vacation—maybe coming back, maybe not. The whole venture seemed to be falling apart. After almost five years of round-the-clock work, I had little hope of success.

And that is when surprise set in. While waiting for the filter parts and frantically working around our other problems, we had another research project under way on advanced carbon nanostructures from the plasma process. One smart PhD student in our lab team discovered a material called graphene in the carbon samples from our test runs. At the time graphene was a new wonder material that many leading researchers wanted to look at. It is made of a single layer of carbon atoms arranged in a hexagonal lattice. It is the thinnest object ever created—a million times thinner than a human hair. It is lightweight and flexible yet also the world's strongest material—200 times stronger than steel. And it conducts electricity faster than most other materials. Some see it as the most important material of the twenty-first century. And all of a sudden we had—rather unexpectedly—a novel process for making a certain ultraclean grade of it.

The lab team successfully completed more new tests on graphene production, which elicited some raised eyebrows from carbon material specialists at the Cambridge Department of Materials Science and Metallurgy. Not much later, a bid for the company arrived on our new chairman's desk. The question was: Should we continue fighting for the hydrogen vision (high-volume, low-price market), or sell the venture as a novel graphene-making process (low-volume, high-price market)? We agreed to enter negotiations. I suggested to the buyers that we could keep the hydrogen solution in a separate spin-off company, and they could buy the graphene process. But that, they said, would be a deal breaker. I would have to let go of the zero-emission transport vision that had initially fired me up.

After six months of negotiations, hundreds of pages of documents, lots of lawyer involvement, and painstaking scrutiny of patents and old liability settlements, the money arrived in our shareholders' accounts. The company was sold. I was out. Our green-growth vision for hydrogen almost succeeded—but failed miserably at the very last step. The potential for even greater healthy growth solutions based on graphene was now in other people's hands, as they—experts on advanced carbon materials—started commercializing the grades of graphene that "our" process can produce in abundance at low cost.

At the time of writing, nobody knows the outcomes of this potential; the novel graphene process might improve batteries, copper conductivity, airplane construction, or other applications. Such applications *could* change the course of this century. Or they might not. Graphene might turn out to be hype, never to have a big-time impact on the markets. As for the hydrogen solution, perhaps those who bought me out will pick it up again later. Having signed a noncompete clause, I cannot.

It's strange to ponder that I've been very close to potentially solving one of the main problems with transport emissions, which make up about 18 percent of global carbon emissions. Of course, there would have been many more problems to solve before getting the process up to scale. Building a million plasma reactors or so to supply the world with low-carbon hydrogen on demand is no easy feat. If we were so close, perhaps others somewhere—with more of an Elon Musk in them than I have—can complete the final steps to market diffusion and success. Luckily, electric cars have taken off in recent years, offering another solution for near-zero-emission mobility. Tesla is forging ahead, forcing the other car companies to respond to the competition from high-performance electric mobility.

A few questions, though, stick in my mind: If I was part of an under-the-radar company almost solving one of our biggest collective challenges in a profitable way, how many other under-the-radar radical innovation companies are right now out there? Ninety-nine percent of them may fail. Maybe some will, like mine, find success in a radically different market. You never know how innovation will occur. The threads of entrepreneurship are ungovernable, mercurial. But with thousands and thousands of these enthusiastic innovators, is progress not inevitable? Or does the diffusion of their small, smart products happen too slowly relative to our large, global problems? Does it matter if we have fancy solutions if we're too slow to scale, invest, and put them to overall use?

THE MAGICAL FACTORY: ONE TOMATO, TWO TOMATOES

One of GasPlas's Norwegian industrial partners, Lindum, has taken biogas further without the final hydrogen steps. Headed by the innovative CEO Paal Smits, the regional waste treatment company takes municipal wet waste and farm manure, and runs it through huge biogas digesters, which look like enormous barrels with rooftops. In these digesters billions of anaerobic bacteria transform all that liquidized organic waste into biogas, which bubbles up to the top. Biogas consists of mostly biomethane and some CO_2 and is highly combustible. Hence, it's a useful energy source and a perfect substitute for fossil natural gas. When upgraded and compressed, this biomethane can replace fossil fuels for heating or replace diesel in trucks, buses, or ships that have suitable gas-combustion engines.

What remains from the biodigester process, after the sales of biomethane, is mostly CO_2 and the nutrient-rich effluent. The effluent sludge contains all the nutrients of the waste except carbon, and is a balanced fertilizer—replacing fossil-based mineral fertilizers at neighboring farms. And recently, Lindum has even started reusing the CO_2 in a novel way: They pump it into adjacent greenhouses, along with some of the nutrient-rich effluent, for hydroponic plant growth. Suddenly you get an almost closed-loop healthy system: A rotten tomato is turned into not only climate-neutral biomethane that fuels public transport but also nourishes new tomatoes in the greenhouse. One tomato becomes two. The associated CO_2 from the biogas is reabsorbed in the photosynthesis of the briskly growing new tomato plants in the greenhouse.

Lindum's prize-winning plant at Greve, in Vestfold County in Norway, has thus been dubbed the "magical factory."[14] The more waste they run through the magical factory, the less waste enters nature, and the less CO_2 enters the air. It is a waste system leading to climate-positive flows. As more biogas systems are built and their by-products are used as fertilizers and return organic carbon to the soils, the better for both climate and Earth, as new research documents.[15] It's a truly regenerative type of economic growth, generating lots of local jobs as this growth model spreads.

HEALTHY GROWTH CASES IN ALL FIVE END-USER SECTORS

I've briefly told the stories of a handful of the entrepreneurs realizing healthy growth in their enterprises: Toast Ale, Solar Sisters, Trine, Interface, Powerhouse, GasPlas, and Lindum. This may seem like a haphazard bouquet, but there is a logic behind the selection. Each case illustrates healthy growth opportunity in each of the main polluting end-user sectors of society: food, buildings, industry, transport, and waste.

In each sector there are numerous breakthrough innovations available and people yearning to put them into broader use. Toast Ale demonstrates how to start fixing both food and waste; Lindum exemplifies turning waste into food and transport fuel; Powerhouse eliminates the footprint of existing and new buildings; Solar Sister makes electricity affordable to people in buildings without grid connection. Trine finances projects with highly beneficial social and climate impacts. Mobility companies, whether as small as GasPlas or as huge as Tesla, focus on new opportunities for cleaner, safer transport. Interface shows how industry and manufacturing can transform from extractive robbers of the Earth into net-positive factory forests, all while profiting.

Happily, there are hundreds of solutions, examples and success stories out there, well documented by such initiatives as Project Drawdown, the GlobalOpportunityExplorer, and others.[16] These people and the fascinating stories behind them need more tellers and broader audiences. Unfortunately, most media focus instead overwhelmingly on the dark C's: catastrophes, corruption, crimes, and crises.[17] Companies, NGOs, cities, and municipalities need help getting their sustainability stories out there: The narrative powers of breakthrough examples are what can inspire successful transitions.[18] Stories become self-sustaining when told and retold, sometimes even self-fulfilling. That's why a sense of purpose and enthusiasm when speaking about them is so critical.

The new version of the archetypal hero's journey, that moves these people goes something like this: The industrial economy of the 1900s treated people and planet like they were there for the grabbing. But the old rule of King Industry is now at way's end. The monsters of global climate weirding, the gruesome giants of toxic waste, and the Godzillas of ocean pollution are all knocking at the gate, threatening mayhem. The King is sick or dying. His advisors are self-absorbed and corrupt. Now brave businesses and innovators must heed the call to adventure, cross the threshold into the twilight zone, face the monsters, and transform themselves to avoid future risks and to rescue society at the same time. Yes, that's risky. But the new heroes may discover the four special powers now at their disposal: *redesigning* consumption, which could do away with waste and pollution at the end user. *Renewable* power, which is becoming abundantly available from sun and wind. *Recycling*, repairing and upcycling, which could keep materials in use for longer. And, finally, *regenerating* the soils, the rivers, the forests, and the oceans, the reinvestment in natural capitals. Are these four magic R's sufficient to slay the monsters and win the prize?

The prize—the lost grail—must be fought for and returned if the Good Kingdom is to regain its health. Anyone and everyone who hears the call is asked to rise to the occasion, answering yes to the grand challenge.

I'm continually surprised and moved by the quantity and quality of people now rising with such a powerful sense of purpose. My impression from speaking with the (often young) heroines and heroes over the years, from attending countless conferences and humbly pursuing my own startup attempts among them, is that they have a very keen sense of purpose. It shines through in their eyes. They are without exception truly enthusiastic. Meeting them really touches and enlivens my heart. You can sense it in the way their hands move. And in the torrent of words running through them, as if overcome with eagerness to share their findings with the world. The more I engage with them, the larger a sense of grounded hope grows in me.

Without the new stories coming alive in the hearts of hundreds of millions, there is little chance of moving markets, business, and politics fast enough. We are at a crossroads where the enormity of the challenge ahead could paralyze us into complacency. This book is my contribution to serve and tell those stories, economic storytelling for the twenty-first century.

Which will win in the coming decades: innovation or inertia? My vote is on the thousands of initiatives focused on resource productivity and circular system redesign that will, I believe, eventually create a grand wave of disruptive innovations, rebuilding our economic system. This wave will reinvent growth itself—causing a creative destruction of our old business-as-usual growth model and making vast new profits, jobs, and wealth possible.

3 HOW ECONOMIES ARE TRANSFORMED: THROUGH GRAND WAVES OF INNOVATIONS

Have you ever flown into a large city late in the evening, glanced out the window, and gazed in awe at the profusion of visible lights? There are long yellow lines of streetlights. Miles and miles of red and yellow streaks made by white headlights and red taillights maneuvering in tight traffic. Innumerable high-rises with rows upon rows of lit windows. Lights beaming from bridges and airports and so much more. It's a vast, awe-inspiring sight, and a unique vantage point on the magnificence of our late-modern cities, so recently constructed.

It's also a sight that can give one a sinking feeling—as it did to me one night when flying over the sprawling metropolis of Los Angeles. I found myself astounded at the magnitude of modern development, but also acutely aware of its source. Most street and building lights in the world's cities come from concealed fires of fossil fuels driving turbines to conjure electricity of which only a small fraction became the morass of lights below. Keeping all those lights ablaze came with a price. I'd just finished reading some news reports about politicians promising to reignite growth and deliver energy security. They were arguing for doling out even more drilling rights to huge fossil-fuel corporations, and for removing regulations that constrained their expansion. There was no other way, they argued, to meet the energy needs of the world we have created. At a time when the climate impacts of burning fossil fuels had never been clearer, the world seemed desperate to stick to its old ways. I felt low and dejected, as the plane sank toward the ground, by the thought that we modern humans are in for a societal and economic crash landing. And pretty soon.

I ticked through the litany of reasons why collapse is approaching, and fast. There was the hesitance to step away from the huge oil infrastructure we have built—with its enormous rigs, pipelines, and storage systems, its mile-deep wells and huge tanker fleets. There were coal mines, coal trains, enormous power plants, hundreds of millions of cars with combustion engines, a petroleum lobby with endless funding and well-trained lobbyists, short-term behavior by corporations, sold-out politicians, and arcane congressional procedures. On top of that, in the climate system, self-reinforcing loops were kicking in as tundra and Arctic sea ice melted. Combine this with the growth in global population—and the wealth, consumption, and material extraction fossil fuels enable—and our fate points in one direction: overshoot of planetary boundaries and then collapse. It's a juggernaut. And there's no driver in control who can step on the brakes. Or he's maybe drunk. Or stark raving mad.

If you are aware of societal, economic, and environmental trends, you've no doubt ticked through the very same list and struggled to find hope in a different way forward. We are often told that the way forward will come to us through innovation. Do we believe it? To answer that question and see our own time in perspective, let's look back at the nature of waves of innovation throughout modern history.

IN THE MENTAL GRIP OF THE 1900s GRAY GROWTH STORY

It's hard for those of us who live in richer countries to even imagine how people thought and lived before the industrial revolution, just a few hundred years ago. All those who live in richer countries today take washing machines, plumbing, hot showers, double-paned windows, penicillin, contraception, dentistry, smartphones, the internet, and planes simply for granted. But just around thirty years ago, even the words *internet* and *World Wide Web* were nonexistent in the wider society.

Innovations are often incredibly hard to imagine before they arrive. And they seem so natural, ordinary, commonsensical after they've spread and saturated our society. The innovations themselves begin to change our worldview, then our world. First, the mind invents technology. Then the technology transforms our language by introducing new metaphors and words. Finally, language shapes our minds and influences how we think. Then, how we think determines what we find natural and what we do. Winston Churchill, speaking about rebuilding London after the Second World War, said it well: "We shape our buildings, and afterwards our buildings shape us."[1] Call it cultural or mental evolution.[2]

Working with data, for instance, on punch cards in the mainframe-computer age of the 1970s was worlds apart from using the computer mouse and Windows screen in the PC networks age of the 1990s. And the PC age is now receding while cloud servers, Wi-Fi, smartphones, and tablets are pervasive. When everyone has smartphones with touch screens, it's hard to remember the days of Nokia button phones or, even more weirdly, the landline phones with circular dial pads. Today, my teenage kids are hardly capable of imagining what a day without Messenger or Snapchat on their smartphones would be like. The tools we use to think shape the ways in which we think, as MIT professor Sherry Turkle noted.[3]

Before such shifts happen, they are almost unthinkable. In hindsight, they look inevitable. And the past starts to look strange, dusty, and outdated. Can you imagine how people managed to get around and meet each other in cities before they had a smartphone?

Becoming aware of how limited our untrained mind is when imagining a different world is fundamental to forming a new narrative of the future. This is core to the discipline of scenario thinking: Without good stories to help us envision something very different from the present, we humans are easily stuck in our conventional mental programming. And today, most people are currently still in the grip of the dominant industrial growth story of the late twentieth century.

In most sectors of society—including many corporations, lobbying organizations, public officials, financial and regulatory institutions, and even much of the business media—people prefer not to change. Let's stick to what we know, they say, stay true to our "core business" and "core competencies." They try to sound McKinsey-like and look tough-minded, like economic champions carrying on our proud traditions of good profit. Like hard-core business guys. We have all seen this ubiquitous resistance to change masquerading as business strategy. It is the immense institutional inertia at the core of business as usual. People like to do what they've always done. Budgets for next year are just like the previous years, only 5 percent larger. We reuse Excel sheets and old designs; we prefer conventional solutions. Psychologists calls it the status quo bias, a strong, automatic emotional bias that prefers the current state of affairs over change.[4] It feels even better if we can reinforce that status quo preference with a "good," seemingly rational explanation. This all-too-human response has been with us for ages. And it is still alive and well today. The grand, gray growth story has a massive grip on our imaginations. It still shaping our society, even as many people speak of sustainability.

So, back to our conundrum: How can we separate ourselves from an ecologically destructive, fossil-fueled economy and its underlying infra-structure? How can or will this vast, seemingly locked-in, path-dependent system change in time? Yes, there is a basket of nice, select, small, net-positive innovation stories like those in chapter 2. But overall the current fossil, food, and financial systems, and the corruption interwoven with them, are so rigid that when thinking realistically it's incredibly hard to imagine that they ever can or will transform. When looking forward, the International Energy Agency (IEA)—like many others that model our future energy scenarios—sees only a slow, gradual, incremental turning of direction, taking decades and decades.[5] Because that's "realistic"—meaning that it aligns very well with the old story.

A NEW WAVE OF DISRUPTIVE INNOVATIONS

And yet, from 2060 or thereabouts, we might look back and view it as inevitable that *it did change*. Radically. And fast, from around 2020.

Looking forward from today, however, it is hard to envision just what will overthrow these now-ingrained, nearly one-hundred-year-old patterns of highways, gas stations, pipelines, coal trains, city power lines, supertank-ers, and exhaust pipes. But it seems clear that our world will be forced away from its primary power sources in the years to come, especially since the fossil-fuel industry has done little new in the last century, mostly sticking to fine-tuning drilling, pumping, piping, processing, and combustion. The trillion-dollar fossil car industry, for instance, will be forced to let go of the combustion engine. Agribusiness will have to let go of fossil fertilizer, excessive meat grazing, Wild West water wasting, and soil-killing practices. The utilities will have to give up their centralized behemoths of coal, fossil gas, and nuclear excesses. The way buildings are conceived and constructed will soon be completely overhauled.

But despite the pains of deep change, the ubiquitous resistance, the status quo bias, and the constant ridicule of radical new by the mainstream, the next innovative wave will be upon us faster than we think. It actually started the day before yesterday. And it will prevail not because of a surge in idealism or a breakthrough in new morals that makes forward-looking politicians to courageously decide to close down the fossil-fuel era.

Our inertia will be swept away because of an unstoppable wave of competitive innovations heading our way. Deep and extensive research has been done on the waves of innovation that have led to major societal shifts since the start of the industrial revolution. Since the first wave of factories and spinning jennies—the first weaving machines of the 1760s—we've seen that time and again a set of fundamentally interconnected innovations change the value-creating logic of the economy.[6] Each wave of technological innovation—lasting forty to seventy years—fundamentally overthrew the old order in a few decades, killing off a lot of old companies, institutions, and infrastructure. Each wave also created vast new domains of opportunity and riches for those whose innovations were ready to ride the swell at just the right time. In other words, each wave of innovation spurred what economist Joseph Schumpeter named "creative destruction."[7]

Before we revisit those waves of innovation, let's recall a familiar story from pre-Victorian England—the story of the classic economist Thomas Malthus. Around 1800, he made some simple calculations that startled him and riled the intellectual world of the British Empire. He observed that population increased exponentially by a certain percent per year when food supplies were abundant. Which means that the more people there are, the more they multiply. But land for agriculture could not grow in the same way. It could only increase linearly by clearing more forest each year. People beget people, but land doesn't beget land. Land is limited, and the best land is cultivated first. The land cultivated later makes less food. On the classic Malthusian chart, one line curves upward (population), but the other (food production) can only—at best—grow straight. Hence, by pure mathematical necessity, a population crash was unavoidable in his calculations.

Sooner or later, the hunger of the new millions and millions would far outstrip any food supplies, he argued. Then, with inevitability, vast famines or pests would wipe out the excessive population. The poor would go first. Thus, the poor masses were doomed—unless they could stop copulating. But the cool-headed, tough-minded Malthus did not allow himself to view a change in sexual behavior as a realistic scenario. That was just an illusory hope against cold, hard mathematics. No way forward except into devastation. According to his observations and calculations, society was inevitability heading toward suffering, famine, pests, violence, and break-down. His conclusion was so dark that it inspired Scottish philosopher Thomas Carlyle to label economics "the dismal science." It stuck. Malthus died in 1834. He did not live to see why his pessimistic prediction was wrong.

But we all know what happened—or didn't. Malthus underestimated the effects of international trade and agricultural productivity. Waves of innovation in ships and railways, tractors and fertilizer production disabled the Malthusian trap and saved the British Empire, though they ushered in their own problems. Today, many see Malthus as a failed economist, a false doomsday prophet, a pessimist, and an unjust, inhumane, and narrow-minded elitist. The label Malthusian is applied to anyone today predicting overshoot and collapse of nature's limits. All he did, though, was to think carefully through certain facts and developments available at the time. The innovations that arrived in the years following his death were unthinkable, unprecedented, hard for him to imagine. None of his contemporary critics pointed to them either. (They were too busy discrediting his character.) Yet in hindsight the innovations appear so natural to us as to have been inevitable.

FIVE HISTORIC WAVES OF INNOVATION

Among the economic historians who have studied waves of innovation and how they impact the economy as well as our society at large we find Nikolai Kondratieff, Joseph Schumpeter, and later Carlota Perez.[8] These are not your typical mainstream economists, but a group that tends to struggle with factoring in the disruptive effects of widespread innovation. Since innovation can be erratic and uneven, it rarely fits well into the smooth, elegant equilibrium models that conventional economists prefer. Rather than clean and balanced equilibriums, the innovation economists who analyze messy history prefer the metaphor of waves: They come, swell, boom, break, and recede. Their timing is hard to predict. And after peaking, they leave nothing dry after washing over the entire shore.

As these economic historians tell us, there have been five main historic waves of innovation since the beginning of the industrial revolution. First came the wave of innovation enabled by mechanization with machines and factories (1760–1830), followed by the wave enabled by steam, steel, and railways (1830–1900). After that came the onset of industry—the mass production of the assembly line, thanks to innovations in chemistry and electricity (1900–1970). The fourth wave was brought on by the expansion of oil, aviation, and electronics (1945–2000). The fifth was the digital wave, launched by information-communication technologies such as PCs, the internet, and mobile phones (1985 to the present). A new sixth wave has just begun: a green innovation wave riding on top of digitalization. This sixth techno-economic wave will usher in an era defined by a transition to renewables, radical resource productivity, and circular flows.

What can we glean from the first five waves that will help us understand the unfolding of the sixth, our main story as we move toward 2050? Here's a crash course in innovation history.

Wave One: Mechanization (1760–1830)

Imagine it's 1750 and you're a visionary riding your horse down to the British Parliament in London—the pinnacle of power in the capital of the empire. You unmount the saddle, walk in, and declare to the lords with a shrill voice and charismatic confidence that in just thirty to forty years, one man will manage to weave as much as 100 to 200 skilled laborers do in one day. "And, by the way," you shout, "this will change the structure of the economy forever." After half a second of thought, they declare, "All my eye, what fiddle-faddle!" And throw you out. Everybody knew that this was a ridiculous prediction, since the speed of weaving had been the same for centuries and centuries. Why should it change all of a sudden?

Before machines such as spinning jennies, coal mines, steam engines, and other mechanical wonders reshaped the country, British society was agrarian. Value was made from land, serfs, and rents, or from trade of mainly agricultural products. But with these novel technological possibilities, economic and social change followed. A new class rose beyond the nobility: the capital owners, those who controlled the mills and spinneries, the weaving and mechanical factories. They got super rich and—after a fight—gained political power, too, sometimes ousting the gentry and nobility. The point is: Such techno-economic innovation waves transform the structure of the entire economy. After a period of exponential growth and frenzy, the impacts of the innovations reach everywhere in society. That's why such waves are sometimes called paradigm shifts.[9]

Wave Two: Steel, Steam, and Railways (1830–1900)

Now imagine that you are living in the Norway of 1830 and you have had a vision of the future: You can suddenly envision a world with railways and huge, long trains with tons and tons of cargo speeding through tunnels and forests, over steel bridges, and into great halls in the cities. This vision feels important and significant, and you want to share it, to paint a picture in people's mind of an exciting, new future of mobility. So you mount your horse and head off to the parliament in the country's capital. You eagerly declare to anyone who cares to listen to you that, in thirty to forty years, one man will be able to drive 200 or even 300 horse carriages of cargo in one day—without any horse! A long silence ensues. Then they throw you out. "Impossible! Will never happen!"

Without converting iron to steel, effective rapid railways *are* impossible. Iron tracks are too soft to remain functional and secure for heavier and longer trains. The innovation of steel production made not just railways possible but also a whole new manner of construction, cheap enough also for high-rise buildings and larger ships. Railways then made long-range travel and quick, cheap transport possible. It also made huge volumes of coal available to drive steam engines. Proud and grand shipping companies that for centuries had perfected wooden sailing vessels soon went broke. This not only changed trade but also spilled over to change cities, coasts, settlements, and agricultural production.

In turn, the transformation led to a new class of people and companies that rose to the top of the economic ladder. Business models that scaled up steel and railways to immense size tended to generate centralized, monopolistic, hierarchical powers. The winner took all. Insane riches were made by the steel moguls and the railroad "robber barons." The Rockefellers, Vanderbilts, and Carnegies became the richest people on the planet. Their oversized power and influence extended into business and politics alike.

Wave Three: Industry (1900–1970)

Imagine that you're in Washington, DC, in 1908. You ride your horse down to see the horse-carriage manufacturers, the whip makers, the breeders, the saddle makers and blacksmiths. Then you say: "Horses will always be the way we get around, just as it's been for the last eight thouand years. Safest business in the world. You guys deliver terrific quality. I've seen the future: It will bring us better horses that run ever faster!"

Finally, after all those previous rejections, it feels good to be saying the comforting thing. You're embraced. They pay you one fat consultancy fee. You get two hugs and three cheers for your brilliant analysis.

The problem, of course, was *not* that electric or petroleum automobiles didn't exist at the time. They weren't unheard of or unimaginable. The problem was that an obscure technology—totally unrelated to horses—called the assembly line was coming into existence. And a fellow named Henry Ford had some radical ideas about another future—one that involved business-model innovations that included living wages and car loans. With those, he could accelerate the rollout of cars to people who didn't have the means to finance their shiny new purchase upfront. Economies of scale made car prices fall and fall. Then cars sold by the millions. By 1920, cities had started to ban horses on roads inside city limits.

The then-new mass-production techniques—along with electricity to power pumps, lighting, cooling, and heating—made incredible volumes of consumer goods available at ever-lower costs because resources were abundant and cheap. The combination of cars, trucks, and a flood of new products kicked off modern life as we know it. Economic growth surged, for the most part, through the roaring twenties onward. Again, those companies and owners that controlled the new innovations—the Fords, the Mellons, and later the Waltons, with the advent of Walmart—rose to the top of wealth from mass production and mass retail at the tail end of the wave. This wave of innovations spilled over into all corners of society, changing nearly every sector. The butcher on the corner, the family-owned watchmakers' shop, and mom-and-pop stores started their terminal decline.

For instance, with widespread electricity came widespread food freezers. All mass-produced foods, whether canned, frozen, or vacuum-packed, needed packaging with new coating materials and plastics. With innovative packaging and surging consumer food demand came new food products and the factories to make them, and further innovations in transport, cooling, retail, and marketing.[10] The extension of this innovation wave wasn't complete before the whole economy had been covered by it. A time traveler jumping from 1890 to 1950 would struggle to believe what their eyes were seeing.

The steel and rail barons, once mind-bogglingly rich just a few decades earlier, were now dwarfed by the new industry conglomerates.

Wave Four: Electronics, Television, and Aviation (1945–1990)

Few inventions have spawned as many innovations as the transistor, a semiconductor at the basis of all electronics. From its lowly beginnings in vacuum tubes, it has not just made radios better but has made telephone switchboards and televisions possible. This spawned television companies, telecommunications networks, and news-broadcasting networks. With this, politics changed forever with the introduction of screened speeches, presidential debates, and live news. On-screen likability became more important than issues or content. One of many surprising consequences of electronics flooding the nations was the reshaping of minds through television, as it spread rapidly to almost all households, see figure 3.1.[11]

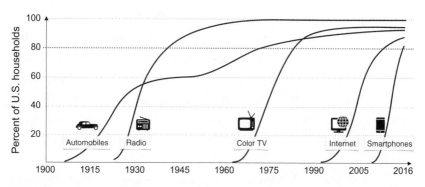

Figure 3.1
A few S-curves from the fourth and the fifth waves, showing how such consumer innovations spread faster over time.

The transistor also made computers possible. By the 1980s, IBM had catapulted to the top rung of the world's most valuable companies. In 1985, it was worth almost three times as much as the second-most valuable company, the petroleum behemoth Exxon from the previous wave.

With advanced electronics and the availability of cheap oil, mass aviation also became feasible. Flying fighter jets and Boeings by mechanical means only wasn't a very attractive option. Large aviation companies sprung up, airports mushroomed, and global trade took off.

Wave Five: The Digital and Internet Wave (1985–Present)

Given the many ways our lives are defined by the internet today, it's difficult to believe that the World Wide Web only got going during the late 1990s. It is still rocking all boats. Once again, we can't really say we saw it coming.

In the late '90s, I was involved with a multiclient scenario-planning project where we looked into the future of digital society. In 1996, many CEOs were still declaring that the internet was a fad. Why would anyone want to purchase their loans, their newspapers, or plane tickets through such a wobbly, cumbersome channel? Back then, for most users going online involved a dial-up connection with slow analog signals over copper wires. There were already hundreds of television and radio channels, and you could reach anyone by landline phone, fax, or post. What would you need an internet for?

Like the other waves, it started in just a few fringe arenas in physics and defense labs off the radar of the general public. The internet erupted commercially during the late '90s, then frenzied and crashed in 2001 and 2008, and has since reached the maturity stage. Now it has spread to all parts of society. "Everything changes" through information technology (IT) and digitization. Today nearly everyone has a mobile smartphone or tablet connected to the internet, something no one had only fifteen years ago.

As with other tech waves, the fifth wave has given us new language. Just as prior waves got us talking about "hold your horses," "driving the highway," "watching telly," or "rebooting computers," our internet age has given us new words that we rarely reflect about, such as *website*, *googling*, and *tweeting*. When our language changes, what we're able to see and do expands too. As do our jobs. Our answers to that old question "What do you do?" change: "I design webpages and facilitate SoMe." That would be social media, by the way. In 1995, both the job itself and the answer would be utter gibberish. With each technical wave also come new social discourses. The two cocreate each other.

And also like other tech waves, the fifth wave has changed the structure and value creation of the overall economy. The petroleum and car companies used to be the world's most valuable corporations, back in the mass manufacturing industry-and-oil wave. Many environmentalists as well as investors still perceive these as large, mighty, stable, profitable entities. They were a core part of most investment portfolios for pension funds and hedge funds alike. But already about thirty years into this fifth digital wave, the shift in value has come about: At the time of writing, the five largest companies by market capitalization in the world are from the fifth wave: Apple, Google, Microsoft, Amazon, and Facebook.[12] Among the companies that owned the fossil-fuel industrial age, only ExxonMobil is large enough to hang on among the very top tiers of global corporations.

The order of things has shifted. And it is not just because Ford, IBM, Kodak, Walmart, and Shell have been shortsighted. It is largely because each new wave fundamentally changes the value-creation logic throughout the economy. Hence deeply ingrained business models "suddenly" shift from leading markets to becoming a drag. A business model is sometimes compared to cell DNA: It hardly changes when the organization is well established. It is deeply embedded in its organizational culture, reproducing itself. When its surrounding industrial ecosystem shifts, rather than rapidly adapting, it loses out to invasive and competing species that suddenly threaten to take over.

Wave Six: Green (2015–2060)

Imagine that tomorrow you drive your SUV to an oil company's annual meeting or a gathering of energy authorities, and tell them that in twenty to thirty years our society can get 100 to 200 times as much mileage and transport work done per barrel of oil burned, if any barrel is burned at all. (By then, burning oil may be seen as stupid, outdated, and useless as a 1990s plastic CD does today.) Perhaps you add that their situation is similar to that of the horse-carriage builders after 1910, punch-card manufacturers in the 1970s, mainframe computer companies of the '80s, the CD music industry in the '90s just before the advent of online music or Kodak before ubiquitous digital photography in the early 2000s.

Will they hug and applaud you for revealing the future? Or will they dismiss you as ridiculous? No, they will want to punch back, to turn the tables on you. After your speech, some journalist may ask, "How did you get here? You drove your exquisite Ford here, didn't you?" Another might chime in, "See! You hypocrite. The car isn't going away anytime soon. And even your phone and clothes are made of petroleum."

When waves of innovation put societies on the cusp of change, those most invested in the old ways rarely grasp the speed with which those ways will become obsolete.

For 200 years, innovators found ingenious ways to improve labor productivity. This was mainly accomplished by having machines (real capital) make people (labor) much more effective per hour. Now we have a world with more than seven billion people, most wanting work. But on an Earth that's restrained in what scientists call sources and sinks—or, in more general terms, raw materials and the air, water, land, and vegetation that can absorb carbon emissions and other pollution—it makes plain economic sense to innovate for optimizing *resource productivity*. The evolution in lighting provides a case in point. The LED bulb gives off the same amount of light as an incandescent bulb but requires just one-tenth of the coal burned in a typical coal-fired power plant. If we then power a smart LED with a motion sensor on wind and solar power rather than power from a coal-fired plant, we can get the lighting we want with at least 99 percent less resource use than the old system.

From 2010 to 2018, average solar panel power costs per kilowatt-hour (kWh) fell nearly 80 percent, becoming cheaper than fossil fuel most places. And solar modules costs dropped over 90 percent from 2010 to 2020.[13] It's truly dramatic—a solar energy revolution. As we saw in chapter 2, resource innovations are happening in buildings, foods, transportation, and industry. All of these are converging into the next (sixth) wave of disruptive innovation. And the wave is clearly on its way. Even mainstream players like the World Economic Forum are now spotting what it calls an "innovation tsunami" that "has the potential to wash over the world's energy systems." In a 2018 report it declared, "Anticipation of this tsunami has been a source of tremendous anxiety. Some firms and industries fear survival. Others foresee riding these powerful waves into new markets."[14]

How will this tsunami impact us when it hits?

Innovation researcher Carlota Perez distinguishes five phases of each wave: eruption, frenzy, turning point, synergy, and maturity.[15] Eruption occurs when there is intense funding for the new technologies, combined with a disdain of old assets. In the frenzy phase, there is a split between real values and the share or paper valuation. One can see inflated expectations in which the value of speculative financial and underlying production capital deviates wildly. Remember the dot-com bubble in 2000, when any startup with a business plan involving e-commerce could find investors? The frenzy stage usually results in a financial bubble, followed by a crash. After the bust, a synergy phase gets a new golden age going again, followed by a coherent growth stage in which production, employment, and share value realign. Finally, at the maturity stage innovations reach market saturation, and there is less big innovation in main industries but more incremental improvements of the products and services. Hence, the economic margins and rate of return on capital slow down. The wave of innovation has sent its ripples all through the main sectors of industry and society. The stage is set for the next wave.

HOW DO SOCIETIES TRANSFORM: TOP DOWN OR BOTTOM UP?

Global problems like climate disruptions, deforestation, soil degradation, and nitrogen overload will never be solved by a united, top-down government approach. We will never see all countries instituting stringent pollution regulations and globally harmonized carbon prices over all regions of Earth. Rather, millions of smart people working from the bottom up will improve their and our lives by wasting less. Gradually, as people encounter more frequent examples of others making more money from fixing resource wastefulness, they will understand that it is a smart strategy. Seeing is believing. Only later, after enough examples have accumulated and this idea, and the companies employing it, have become widespread, will mainstream politicians follow suit. Politicians hardly ever lead by going in front. But they can then, if supported by industry and voters, start raising the bar for everyone, the late adopters, through regulations or taxes.

The guiding mantra of the sixth wave is not only "Less is more" but "More with less." New innovations can make it possible for nine billion people to live well on one planet by 2050.[16] They can enjoy more goods with fewer bads. And this is where the sixth wave and healthy growth intersect.

So will "technology" save us? Is it sufficient that we somehow push new technology and innovation? This question is surprisingly divisive. On the one side, we've got the techno-optimists and ecomodernists who view technology as a sole savior. It will fix it for us. Technological change is always accelerating, they argue: "Technology is the future!" Some, like inventor and tech guru Ray Kurzweil, believe in a soon arriving "singularity point," where technological change gets so exponentially rapid that more or less everything changes at once. On the other hand, we find the modern

Luddites, who insist that you can't fix a problem created by technology with even more technology. Some stress the need for appropriate technology in Ernst Schumacher's small-is-beautiful tradition, and prefer local, low-tech, slow food, natural, all organic, analog stuff. I love hanging out with both camps, and I love to pack my stuff and travel between them.

Yet a pure bottom-up thrust from new and better technology rarely makes it through all the layers of obstacles that are built into previous institutions. Some guilds from the Middle Ages are still around. And new trains can't run without nationally coordinated railroads, bridges, tunnels, and the maintenance of them. Cars get nowhere without open roads and interstate highways. Radios and televisions don't work without clear agreement on the broadband spectrum. GPS can't work without satellites. Advances and breakthroughs in transistors, microchips, GPS, the internet, or even fracking don't happen without large research grants. And solar panel technology needed decades of R&D funding and publicly guaranteed power prices in order to grow enough to really drop its costs. The point is that *all technologies* and *all markets* exist in historic, social contexts inside governmental regulations, which are run according to deeply ingrained cultural values and ideals.

So, transformations may get momentum from the intense market pressure of new technology, like electric cars or autonomous driving. But without government response and legal guidance, innovations are often unable to break through into the mainstream. The speed of diffusion is too slow. Incumbents and their old systems are too entrenched. There are too many hurdles and too much red tape. The quality of new and unproven tech may be oversold, leaving people feeling duped (as with the fake "low emission" diesel engines that led to the 2015 Volkswagen scandal). Therefore, any innovation will accelerate better if it can surf on underlying societal driving forces as well as thoughtful regulation in order to gain market momentum. Markets are always embedded in society, reflecting long-lived power structures.[17] New innovation waves need sufficient height to break through the historic, established obstacles and infrastructure. Individual states and governments can hamper or enhance this, but rarely stop it altogether.

To answer this section's main question in a nutshell: Societies transform by first seeing a swelling wave of new, emergent, bottom-up initiatives, which are initially met with resistance, then political acceptance and top-down institutional reform. Bottom up first. And when the S-shape rises high enough, even the mainstream politicians will finally follow suit and surf the wave. Technology by itself can rarely do it. But when the societal system embraces the technology and starts integrating it into existing structures, then it may contribute to society's main goals. The ripple impacts may then come as a sudden surprise to those accustomed to the status quo.

DRIVING FORCES OF THE SIXTH WAVE

Why will a sixth wave build into a massive change? The coming sixth wave of innovation builds momentum and height from at least four converging gales: radical end-user efficiency solutions, rapidly falling costs for renewable energy, circular material flows, and, finally, rising risks and costs of new fossil investments.

When the first wave of mechanical innovations got going, humanity and its total economy were pretty small compared to the vastness of Earth's uncharted oceans, extensive forests, unexplored mountain ranges, plentiful rivers, and wild animal herds. Machines and capital, on the other hand, were new and scarce, and skilled labor was relatively expensive. It made economic sense to grab whatever bountiful resources could be taken for free while improving labor productivity by creating ever better machines. The more each worker could produce per hour, the more could be sold. And then profits would rise. That logic continued for about two hundred years. During all that time economic decision makers put less priority on resource productivity. Resources were relatively cheap and abundant. And there was little to pay for the pollution.

But exponential growth is devilish. Slow and imperceptible at first, then, after several doublings in human population, the footprint of the whole human economy, once dwarfed by the scale of Earth's bounty, "suddenly" became large relative to the size of the planet's sources and sinks. There were billions more of us, and 91 percent of everything that we took from nature, from fossil energy to biomass, became waste anywhere from a few minutes to a few months after its use.[18]

Only in the last decade or so have we seen substantial improvements in energy and material productivity in some richer regions, like North America.[19] Radical resource productivity means, quite simply, much more with less. More well-being and more human needs met—with just a fraction of the conventional twentieth-century resource use, and reduced footprints on nature.

The logic of sixth wave innovations relies on shifting away from our 250-year focus on labor productivity and toward a new focus on resource productivity. It makes increasing sense to achieve this by raising resource efficiency radically, by a factor of 4, 5, 10, and in some applications even 100 and more. Resource efficiency means making the exact same useful benefit or service with less resource input.[20] The good news is that radical resource efficiency is not just feasible but also profitable. Wastefulness isn't just destructive to nature; it is also bad design and bad business.

Driving Force Number One: Radical End-User Efficiency

Efficiency, productivity, intensities,[21] waste, sources, resources—all these boring words are becoming sexy in the sixth wave. It's cool to be lean resource-wise while being creative, clean, elegant, and abundant design-wise. It means adding miniscule sensors and smart features and redesigning material flows with the end user's human needs as a guiding star. We can design away pollution, toxics, and waste from the get-go. Thus, we can replace the insanely wasteful patterns of the twentieth century with clean system redesign. Better lives with less waste and pollution as incomes per person rise.

Need indoor light? Rather than burning coal to boil water to make power to run through a grid to make light by heating a thin wire at the desired location, just design the building to let daylight in during the daytime. In the dark hours, let sensors determine if someone is around. If so, automatically optimize the LED lighting.

Need to go somewhere? Rather than owning your own car, get a ride share in a taxi or on a bus or hop on your own favorite e-bike. Or, better yet, why even own a bike when you can bike share? You can be guided through the best options by integrated mobility-as-a-service apps.

Need clothes? Rather than purchasing a dress made from conventional pesticide-sprayed cotton, you can rent one from a clothes-sharing company. Or have some AI push an alert to you when fancy, suitable garments from upcycled fiber are available for hire, lease, or purchase near you.

Need a shower? Rather than heating your water with coal-based electricity or gas or oil, use solar heating from your rooftop and store it in well-insulated tanks until you're ready.

Need the restroom? Why on earth use gallons of clean, drinkable water? Low-flush and no-water toilets are gaining market share, saving both water and costs.

One research project identified twenty-one major upcoming consumer innovations that have disruptive potential. Seven are in mobility: e-bikes, bike sharing, taxi-buses,[22] ridesharing, car sharing, mobility as a service, and better telepresence. Seven are in the power domain: photovoltaics like solar rooftop with storage, peer-to-peer electricity (selling to your neighbor), vehicle to grid (selling from the car battery back to the grid when demand is high), disaggregated feedback on your consumption (to lighting, washing, cooling, etc.), time-of-use pricing, managing demand (of washing or heating) according to load, and more energy service companies who will optimize your home consumption in exchange for a fixed fee. And seven are in smarter consumption: peer-to-peer goods (sharing tools, sports gear, etc.), home sharing (like Airbnb), the Internet of Things at home, smarter appliances, prefabricated retrofits with click-on insulation plates, smarter and self-learning homes, and heat pumps. Now, that's an abundant wave of innovations coming toward us, enabling a Low Energy Demand (LED) future with better lives.[23]

In the twentieth century, human needs were met through cheap, abundant energy, most of it from fossil fuels. We turned on the lighting thanks to coal and gas. To get heat we burned oil and gas. Even our food was grown with fertilizer made from natural gas! We used oil and gas to make plastic packaging and clothes. And, of course, in these early 2000s, we still do. But it is the twentieth century that will be labeled the "fossil century" of civilization by future historians. Going forward into the twenty-first, more and more human demands will be met more efficiently without using any oil or gas. That's right: zilch, zero.

When houses are built with temperature, light, and energy optimization in mind, they don't need much external energy. Conventional boilers and heating systems can be replaced with smart, passive houses that are so well insulated they hardly need heating or cooling. The small remaining needs can be met with solar-powered heat pumps. You can then cut out the natural gas supply. You don't need a heating system or to install power-hungry air conditioners. Ventilation systems can be downscaled and cheaper. Buildings become net-positive, meaning they produce more energy than they consume over their lifetime. Being connected to the grid will still be required, in order to trade power with other buildings and the utilities. But on most days buildings might even sell more power than they buy, becoming small power stations while also charging any electric cars connected to their circuits. The electric cars' batteries may help stabilize the grid, too, optimizing the timing of the power buys so that one can buy power at low cost and sell it when the price is highest.

And so it goes. For all human needs, there are radically efficient end-user solutions ready to go big time. There are a great many startups, scale-ups, and innovations within large corporations aiming to commercialize these opportunities. They are coming to a marketplace close to you in the near future. Among the new players we find Whim, based in Helsinki, the home of Nokia. Whim is an app offering mobility-as-a-service (MaaS) for all your transport needs, whether public transport, bike share, taxis, and affordable rental cars at a fixed monthly rate. Or the San Diego–based vehicle-to-grid company Nuvve, which can help you make money off your electric vehicle even when it's parked by trading power storage to and from the grid. There's Spinnova, which offers textiles spun from forest pulp and agricultural wastes. The fibers are warmer than wool and stronger than cotton; there are no toxics in production, and the materials are fully recyclable and compostable. And there's DesertControl, which injects nanostructured liquid clay to regenerate arid or degraded soils. Whether in lawns, gardens, parks, golf courses, or fields, it can increase the soil's water retention, organic carbon, and microbacterial capacity. It can drastically improve soil health while cutting fertilizer and freshwater irrigation, both in urban and rural regions. All of these companies—and many others—are working every day to scale up to serve global needs as quickly as they can.

Such radical resource-productivity solutions are ready to sweep the floor of old products and services simply because they offer superior value. Not just because they are greener, but because they are better—for you, for ecosystems, and for profits.

Driving Force Number Two: The Rise and Rise of Renewable Energy

With end-user efficiency we will need much *less* energy and resources when serving more human needs. With net-positive houses and electric battery cars, gas and oil demand is being substituted. But what will really kill fossil fuels in the coming decades are, of course, the continuing improvements in solar and wind power with energy storage. Each time the number of installed solar capacity doubles, the module price per watt drops around 26 percent.[24] This phenomenon contrasts starkly with the increasing long-term risks and investment cost trends in expanding new gas, oil, or nuclear capacity.[25]

Since the turn of the twenty-first century, we've doubled global solar capacity around nine times (from 2 GW in 2000 to ~750 GW in 2020), which has led average prices to drop more than 90 percent since 2000. To a large extent this price drop was initially kicked off by Germany's generous feed-in tariffs in the early 2000s. Then China's huge overinvestments in capacity crashed the price again since 2010. "Wind and solar were in 2019 cheapest across more than two-thirds of the world. By 2030 they undercut commissioned coal and gas almost everywhere."[26]

Every day in 2016, around 500,000 solar panels were set up somewhere. Just two years later, it was around one million panels per day. Producers are gearing up to increase their capacity to install even more. In a few years, there will be many millions of new panels installed every day. And efficiencies are wrenched out of each silicon ingot slice and square inch of solar film. Smarter ways of making materials for solar panels are being commercialized. It is inevitable that the overall costs will continue to decrease while economies of scale grow.

Solar conversion efficiency is also being improved year by year. On premium solar panels it's now above 21 percent, which means that 21 percent of the energy in the sunshine comes out as electricity. For comparison, the efficiency of photosynthesis is only around 3 to 6 percent.[27] Some competing solar power systems have now started to use both sides of the panel, while others use mirrors to concentrate the sun onto the solar panels. This can then achieve even higher conversion rates, particularly if co-generating heat. They can run 40 percent cheaper than conventional solar panels due to cheaper mirrors and fewer panels of higher efficiency. Other companies are bringing even cheaper panels to market based on new production methods and materials, such as perovskite.[28] We've certainly not seen the end of solar and wind power innovations. We're in the midst of a techno-economical paradigm shift, to use Carlota Perez's language.

Wind has shown the same type of falling costs as solar, if not as quickly. Battery costs, however, seem to be falling as quickly as solar power, an average of 24 percent per year since 2010.[29] The main question affecting the pace of solar and wind power growth is how much investment there will be in new annual capacity installation. And the volume of investment is still influenced by policy and tax credits. But whether tax credits are increased or slashed in any one country doesn't change the global dynamics: The rise of solar and wind power competitiveness is unstoppable. If one country strangles support in policy or with taxes, others will take the opportunity and run with it. The key issue is how rapidly the global annual installed new capacity grows each year. Annual solar and wind installations have grown, in fits and starts, at rates of 0–50 percent per year since 2010. But with just on average 10 percent annual growth, the new installed capacity doubles every seven years. And once installed, each panel provides power for around 40 years or more, almost for free, as they require minimal maintenance.

The main blows to demand for oil will of course come from batteries. As battery-electric vehicles (often called EV or BEV) grow their market share, demand for gasoline falls. In 2015 the one millionth EV car hit the road. The next one million cars took about 15 months. During 2019, more than 2.2 million new light vehicles with plugs were sold.[30] When you make a million of something, prices really start to come down, and quality can go up. The combustion engine vehicle industry has had a century of leeway. Producing more than a billion cars, it has had ample time to cut costs. But the fossil car is now toast as it can't compete with the simplicity, safety, performance, cleanness, and cost of ownership of the electric motor running on abundant renewable supply.[31] Now it's only a question of how quickly consumers will stop buying them and/or governments will ban them.

It's easy to get lost in the exciting frontier of solar, wind, and battery development. So much is happening during any given month somewhere in the world that no person can really manage to keep an overview. This reality should put terror into the hearts of coal, oil and gas managers. Their business model is being undermined. There will still be a few more decades when there is substantial petroleum demand. But as we pass 2020, there will be no more *strong demand growth*. The fossil-fuel markets will be in mid- to long-term decline due to squeezed profit margins, from slower end-user demand combined with rapidly rising competitive renewables.

Driving Force Number Three: Circular Material Flows

After radical end-user efficiency and renewable energy with storage, the third large shift comes to linear material flows. Incredible amounts of materials are extracted from nature each year in order to serve the needs and wants of people. But most of that ends up as waste after one or no use. This causes deforestation, acidification, dead sea areas with too many nutrients, soil loss, plastics in the ocean, and so on. Very little of the extracted resources are kept in use for long, nor are they returned for a new cycle after use. Research estimates that the economy is on average only 9 percent circular—meaning that 91 percent of extracted resources end up as waste after one or no human use per year.[32]

Four main categories of materials flow through the economy: fossil fuels, minerals, metals, and biomass. The first two driving forces will reduce and eventually eliminate the flow of fossil energy. This third driving force curtails the need to extract more minerals, metals, and biomass by transforming today's linear growth of take-make-waste into circular flows.

In the coming decades, humans will increasingly learn to emulate nature in this regard. In 2020, the European Commission has launched an ambitious Circular Economy Action Plan, which encompasses not just waste but the entire material flow of minerals, metals, plastics and biomass. The plan aims to create a win-win situation through circular economy measures: Savings of €600 ($650) billion for EU businesses, equivalent to 8 percent of their annual turnover, while creating 700,000 extra jobs and reducing carbon emissions by 450 million tons per year by 2030.[33]

The biggest category of minerals humans use is concrete. And the largest footprint comes from extracting limestone and converting that to portland cement. After water, concrete is the most widely used substance on the planet. Since 2003, China has poured more cement every three years than the US managed in the entire twentieth century.[34] Continued urbanization in the coming decades will undoubtedly require more concrete. But concrete recycling is an increasingly common method of using the rubble after demolition or renovation. Previously, concrete ended up in landfills. But reuse of concrete by cutting or by crushing and recycling has a number of benefits. It can keep construction costs down, cut transport, soak up CO_2 in recycling the concrete,[35] and comply with environmental laws and growing awareness. Just outside Oslo, Norway, a very large shopping mall is to be demolished, and the developer is reusing the concrete and other materials for the new suburban sustainable village construction on the same site. Otherwise, they would have had to buy all that as fresh materials. The reuse saves around $160 million in costs.[36] This illustrates the essence of the circular material flow driver: cutting waste, costs, and extraction by reusing and upcycling materials in thoughtful ways.

Leading companies are finally getting around to designing circular material flows for their products; Apple has introduced the robot Daisy, which can pull apart 1.2 million iPhones a year, or 200 per hour. Apple sends batteries recovered by Daisy upstream in its supply chain. They are then combined with scrap from select manufacturing sites, and, for the first time, cobalt recovered through this process is now being used to make brand-new batteries—a true closed loop for this precious material, says Lisa Jackson, Apple's vice president of environment, policy and social initiatives.[37] Electric car manufacturers are getting in place similar arrangements for the huge amounts of car batteries that are currently hitting the roads.

A more circular economy can reduce both virgin extraction and CO_2 emissions. Increasing the reuse, recycling, and upcycling of just the four value chains of plastics, steel, aluminum, and cement could almost halve the emissions from these sectors globally by 2050 (down from 9.3 to 5.6 gigatons of CO_2 per year).[38] It is potentially the second biggest lever for CO_2 emissions reduction after clean electrification.[39]

We can measure the degree of circularity as the share of cycled materials relative to the total material throughput in tons per year. As this share of cycled materials grows from 9 percent today to possibly over 50 percent by 2050, the economy's environmental impact will shrink accordingly. A decisive point that enables this is that the stocks of material resources (in capital such as buildings, machines, cars, and infrastructure) are around ten times larger than the annual extractions and still building up.[40] As the economy matures out of the industrial age's linear growth model, and better recycling designs and methods are scaled up, huge stocks of already-used materials are becoming available as "new deposits." Existing stocks can be leveraged at low cost, rather than always going for more extraction from nature. The use of digital technologies, such as sensors, AI, and blockchain, in transparent, circular supply chains can significantly accelerate this driver.[41]

A key step on the road to higher degree of circularity is to eliminate the use of fossil fuels, which are the ultimate linear resource, from extraction to only one-time combustion. This is where the fourth driving force interacts with the three we've examined so far.

Driving Force Number Four: The Rising Risks and Costs of Oil and Gas Investments

Cities are economic powerhouses. More than 80 percent of global GDP is made in cities. In recent history, most cities have run almost entirely on coal, oil, and gas. Vast volumes of black, gooey and gassy stuff enter them and "disappear" into the air. Seventy percent of all human emissions come from cities, though they cover just a tiny fraction of Earth's surface.

Where does all the energy come from? From far away, deep down, and ancient sources. A hundred years ago, the oil and gas that humans started to use wasn't that hard to find. In 1912 you could take a small drill, make a hole in the ground in Texas, and oil would gush forth. Using just one barrel to drill, you could bring 100 barrels of oil to the cities. People rushed to find more. Some became insanely rich by hitting and owning the right location. The search became more sophisticated as better acoustics enabled petrogeologists to find more and more reserves. The cheapest, easiest oil and gas supplies were explored and extracted first. Then the sources that were farther away, deeper down, and harder to get at were tapped. But even the nearest, easiest-to-exploit oil fields lost pressure after a while. More pumping became needed. Fields quickly become less productive.

By the 1950s, the average energy return on energy invested (EROI) in the US was down to 1:50 (from 1:100 in 1912). Today, EROI in US conventional oil fields is down to about 1:9, and the shale oil wells of Bakken and similar oil fields may yield only 1:4. An immense amount of energy is now combusted to drill, press, boil steam, and inject and break the shale far under the ground. That means some places now burn one barrel of oil (or the equivalent of gas) to produce as little as four new barrels.[42] The yields decline quickly for each new shale well over time. It may drop more than 60 percent during a single year.[43] Then they must drill and steam and frack again. Economists call this a game of decreasing returns.

Yes, new technology and petroleum innovations may cut costs, and supply may increase when adding more rigs. Innovations within the fracking industries, including horizontal drilling, have increased oil and gas supply from both conventional and unconventional sources, such as tight oil or tar sands or heavy oil fields. What's common to all extraction, however, is that more energy is used on average for each unit of energy returned as time passes. And the later, newer fields and finds are on average more difficult to exploit and bring to market than the first, which are now mostly depleted.

This dynamic is playing out on a global level. The average EROI for petroleum has been sinking year by year since 2000, irrespective of high or low oil price, while solar and wind are seeing improving EROI.[44] The question is not if renewables will overtake fossils but how quickly. There is a race between petroleum exploration innovation and the inevitable decline in the geological availability of remaining easy reserves. There are huge remaining reserves, but they are found in the ultra-deep sea, the Arctic, in heavy oils, tight rock formations, or in distant and politically unstable areas far away from the cities where the hungry cars and power stations want to explode and combust and burn all that dark, gooey ancient carbon-rich stuff.

Let's follow oil from well to wheel: To get it into a car driving through the city, oil is pumped, piped, stored, shipped or railed, refined, piped and trucked again, tanked, pumped, driven around, and, finally, in a microsecond, explosively combusted! Gone. The driver must go back for more. It's an incredibly complex, long linear supply chain. Some may call it a value chain. But it seems more like a *waste* chain, since its external social costs are not figured in.[45]

Why is the declining trend of EROI in oil important to understand? Because it guarantees that there is no chance that this industry will get to a point where suddenly large, new volumes of cheap, light, and sweet crude will appear again. Ever more rarely an oil company will have a lucky strike and find some easy oil next to existing fields. One example is the Norwegian offshore Johan Sverdrup oil field, the biggest in western Europe, which started producing in 2020. It will produce a so-called "sweet crude" at relatively low cost for a few decades. But this type of new find is rarer and rarer.[46] Yet there is more than enough oil in remaining reserves, but they demand ever more energy to continue producing. The oil industry is now trying to squeeze the last barrels out of mature basins at as low a cost as possible.

But they better hurry up if they want to stay in business. What will become abundantly clear as one understands more of the sixth wave and the driving currents behind it is that there is limited time left for profitably exploring, producing, and selling oil and gas. Why? Because we'll need less and less as resource productivity improves. In OECD countries, for instance, demand has been dropping since the 2010s. China, too, seems to be reaching peak emissions sooner than expected.[47] And we'll need even less in the near future due to fossil-demand destruction, and because cheap renewables with storage are rapidly outcompeting fossil fuels. All the while, there remain many oil producers—like Saudi Arabia, Russia, Iran, Venezuela, and even US shale oil—with big reserves who want to make sure "their" remaining oil is sold before they become "stranded assets." That may happen either because demand declines as renewables take over or because policymakers are successful in curbing oil use due to global heating. As financial climate risks increase, investors are looking elsewhere.[48] Still, oil and gas will eventually be unable to compete against solar or wind energy, which has near-zero operating costs.[49]

When you start to look closer at what we use fossil fuels for, you discover that there are competing substitutes entering markets all over the place. Power, heating, materials, transport, buildings, and industrial processes: each can—and eventually will—be reduced to near-zero emissions, even in the harder-to-abate sectors.[50] The demand for fossil fuels is steadily decreasing in richer countries, and the financial climate risks are becoming clearer, at the same time as new supply is getting harder to extract. The shift does not mainly depend on climate idealism in politics, or because all business suddenly wants to "go green." Certainly, removing the current perverse public subsidy for fossil fuels[51] and adding new government regulations, carbon taxes, and more business responsibility can all accelerate it. But the main drivers will be because the substitutes get better, safer, healthier, more attractive, and—finally—more profitable.[52] We're not fully there yet today, nor tomorrow. But we're not far away either.

The energy system is experiencing the creative disruption Schumpeter described: a full overhaul driven by newer, better, smarter solutions. Yes, old thinking, outdated regulations, lobbyists, corruption, bad risk management, underpriced emissions, and perverse governmental subsidies for fossil fuels can slow it down. But they can't stop it as we move forward in the twenty-first century.

AND WITH THE WAVE, A MINDSET

As we've seen, technology shifts also become mind shifts. So what new language, new perceptions, and new paradigms will the sixth wave usher in? It's possible already now to see the contours of a healthy, clean economy, but we would be defying historical lessons to assume we can predict everything about its shape and scope, or even its ultimate impact. The sixth wave is still far enough away to harbor some mysteries. But it is also close enough to reveal some clues about how to ride it well. That involves, of course, changing the nature of growth itself—getting clear on what we want growth to do for us and for the planet, and whether it can someday be truly healthy.

Either way, the wave is approaching, reinforced by end-use efficiency, cheap renewables, circular designs, and risky fossils. Soon, what seems impossible today—becomes the inevitable. Which means it's time to do everything in our power to steer the impact in a just and regenerative direction. Not only do we need digital disruption of the current system, we need to navigate toward a "good disruption."[53]

II THE GROWTH COMPASS

4 SNATCHING WEALTH FROM THE JAWS OF WASTE: GREENING GROWTH

When Disney released *WALL-E* in 2008, it gave us a visual we hadn't ever seen in a family blockbuster: the mountains of trash we leave behind in our everyday lives. It was impossible not to be heartbroken as we watched the smart, cute little robot tidy up a dusty, gray-yellow, trash-consumed planet Earth in the year 2500, one piece of detritus at a time. Hundreds of years before that, humans had fled the used-up planet for space, where they were now living in huge spaceships, being overly well fed and entertained by virtual reality games. The tiny robot named WALL-E was left alone to clean up the huge mess they left behind.[1]

It's hard to perceive—or even imagine—the extent of real-world wastefulness that the modern growth model of the twentieth century created. Maybe that's why we needed Disney to help us visualize it. But a decade has passed since the movie's release, and despite more frequent images of the Great Pacific Garbage Patch, e-waste piling up in ports, or other heaps of trash with nowhere to go, the wastefulness still remains mostly invisible to us: gigatons hidden in landfills submerged in our oceans; burning in smokestacks that turn them into other hazards, like toxic ash and climate-damaging emissions; or entering our land, water, and bodies as invisible microplastics, dioxins, or persistent engineered nanoparticles. The sheer amount, from nano to giga scale, is indeed turning the entire planet into a trash bin.

Is it even possible to change? After years and years of inadequately addressing a problem we know full well exists, the answer might lie in looking at solutions quite differently.

FOLLOWING THE MATTER: IN BUILDINGS, TRANSPORT, AND FOOD

According to an old adage from 1900s, to get to the bottom of something, you should follow the money. Why do cars and trucks still guzzle fuel and homes still heat with oil and gas? Just look at the political influence of the oil companies, you might say. Why do we put up with pesticide-laden food grown in ways that damage land and water? Just look at the influence of Big Ag.[2] When we're up against powerful forces, change seems far, far from attainable. Even when we think we know the solution, we are loath to begin the battle against so many corporate Goliaths.

But there's another adage more fitting for our age of waste: follow the *matter*. A whole new avenue of study called material flow accounting (MFA) is currently looking into the way materials flow through our economies. Following the money, as conventional accounting has done, is all very well. But following the flow of matter and energy gives a much better view of the *real* economy, that whole physical system underlying the money flow, which—in contrast to money itself—is directly serving up the basis for a good life.

MFA, and the associated life cycle analysis (LCA), shows us the astonishing extent of material wastefulness in modern economies. And that is where it gives us a new, more approachable perspective on how to make system change. By following the flow of matter through our buildings, our transportation, our food system, and our cities, for instance, we arrive at a key insight: It's most effective to start at the end of the value chain, rather than at the beginning, where the extraction of virgin resources—like mining, cutting, or drilling—takes place. Radical resource savings at the end-user side can compound backward, back up the supply chain, to mean better services with ever less extraction at the start. That's why I'll start with reexamining what people actually want and need on the consumer side: the end-user *services*.

That might mean useful lighting, comfortable temperatures at home, the ability to get where we want to get at the right time, or access to nutritious foods. These services increase human living standards and even our subjective well-being. No one has any direct use for a lump of coal or ore.

No one eats electricity. Or CDs or DVDs. There is no utility from air-conditioning empty rooms, or from foods that are thrown away uneaten or that make us unwell. Nor is there any happiness from sitting stuck inside two tons of steel and glass in a crawling, noisy, stinking traffic jam.

All these services—lighting, temperature control, eating, entertainment, mobility—can be provided with much, much less energy and raw materials than today. Does the world need more energy? Well, the world's people will need more *energy services*—but very probably with less raw material extracted from the ground. And the incredibly good news is that renewables, electrification, and radical efficiency may soon start to reduce the total primary energy need, despite increased population, more energy services, and higher standards of living for more people.[3]

In fact, eliminating our ubiquitous material wastefulness by creating new, competitive, and efficient solutions for our fundamental service needs is the greatest business opportunity of the twenty-first century. Moreover, it's an economic pathway that doesn't incite division. If you talk about instituting limits to growth, introducing new regulations, or shutting down polluting industrial jobs, political tensions rise quickly. But *nobody* is (publicly) in favor of wastefulness. Reducing it is, politically speaking, a bipartisan goal—equally important to conservatives and liberals, business and government, investors and environmentalists. Making money and jobs from avoiding, eliminating, and upcycling waste is the ultimate win-win-win solution. The question is no longer *if* it can be done. It is all a matter of *how*: how to scale existing solutions, and how quickly new solutions can move from idea to action. Achieving this scaling up lies at the core of healthy growth.

Let's look at a few examples, all of which show how radical resource productivity is being deployed right now—and how it could be expanded.

Eliminating Energy Waste in Buildings

Buildings are responsible for 55 percent of global electricity demand and cause around 20 percent of all greenhouse gas emissions (and 30 percent of CO_2 emissions).[4] Essentially, *all* of that—the energy used and the emissions produced—should be seen as unnecessary waste. And eliminating that waste should be better viewed as an opportunity for healthy growth by creating more value by cutting operating expenses. Whether erecting a new, state-of-the-art building or refurbishing an old one, nearly all of the energy use and emissions can be eliminated, or at least improved by 90 percent, without prohibitive upfront costs.[5] Investing in these solutions is increasingly profitable. The exact payback time will vary, of course, from location to location. And the financing arrangements and contractual designs between landlords and tenants set up numerous obstacles. Even so, it is a highly effective way to turn waste into wealth.

Buildings provide us with a number of services such as shelter, comfortable temperatures, lighting, and drinking water. Imagine replacing all conventional incandescent and halogen lamps with better daylighting and smart LED bulbs that turn themselves on or off when people are around. This will give the same useful lighting while cutting the need for primary energy such as coal or gas burned by 80 to 99 percent depending on the exact configuration. In 2018, global LED sales outperformed sales of less-efficient fluorescent lamps. They are on track to take over nearly all lighting services by 2030.[6] LEDs are also less costly in the long-term. This trend is accelerating into the 2020s.

Or take safe drinking water. The total energy required for *bottled* water ranges from 1.5 to almost 3 kWh per liter. That includes not only producing and transporting the bottle itself but also the water processing, bottling, sealing, labeling, and refrigeration. In comparison, producing tap water in the building you're already in typically requires about 0.001 kWh per liter including for treatment, filtering, and distribution. Drinking your daily liters from the tap rather than a bottle slashes energy use per liter by 99.9 percent.[7] And there is zero plastic waste. In addition, bottled water costs between 240 and 10,000 times more than tap water. The most common reasons given for drinking bottled water are taste and health concerns, despite the fact that tap water is under more control and restrictions than bottled water in nearly all areas. Using a filtering pitcher or an extra in-house drinking filter could give even better-tasting water without changing the energy picture much.[8] A number of companies that deliver such systems for buildings are scaling up with new business models.[9]

I've got nothing personal against plastic bottles, as PET is a pretty good material if it is kept in the loop. If I use a PET bottle, I try to reuse it multiple times, refilling over and over from the tap. But the annual production of 500 billion one-time plastic bottles per year, or a million bottles every minute,[10] becomes our current mammoth plastic pollution problem. There is a widespread belief that the environmental impacts of bottled water production and consumption are mitigated through recycling practices. However, even if recycling is appropriately undertaken, it only saves one-third of the energy in the production stage.[11] Additionally, the quality of the plastic degrades each time it is recycled, thus limiting the number of times plastic can be recycled.

In the cases of bulbs, bottles, and buildings, simply by switching from one-time use to multiple reuses and finally recycling we can save both money and resources in the order of 90 percent or more. This is what we mean by *radical resource productivity*.

It's not just about lighting and drinking, of course. As the story of the Powerhouse in chapter 2 highlighted, abundant solutions can be integrated into better systems for constructing and operating buildings. They include improving insulation and air-conditioning, optimizing windows, installing heat pumps and passive solar, reusing building materials, and using passive ventilation and heat storage and recapture. Designers are also looking to nature, where nothing is wasted, for inspiration. An iconic high-rise in Harare, Zimbabwe, eliminated traditional air-conditioning by drawing upon lessons learned from termites, whose mounds expertly regulate airflow. Other innovators, too, are looking at ways to make buildings regenerative. That means to fully integrate buildings with their natural surroundings so that their resource flows—from energy to water and green roofs—work with nature rather than against it.

Most buildings have the potential to transition from being huge resource-draining, waste-generating monsters to becoming net-positive over their lifetimes.[12]

Eliminating Fuel Waste in Transportation

Mobility is another sector where waste is ubiquitous. Advertisements for that new petrol-powered BMW or Chevy SUV might give the impression

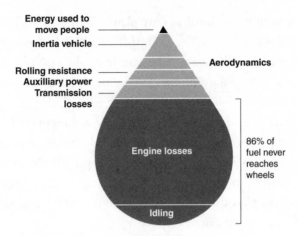

Figure 4.1
Tank-to-wheel energy flow in petrol combustion engine.

that they're the epitome of technological advancement: automatic braking, parking aids, cruising, hi-fi sound. But when looking closer at what we want (the service) relative to what the car consumes, we see again some staggering wastefulness.

In order to bring us from A to B, the car combusts petrol or diesel. Both have amazingly high energy density—around 10 kWh of chemical energy per liter. Thus, every drop is packed with oomph. But how does the car use this energy? Well, first—as with a coal plant—most of it is lost as heat in the engine. It goes out the exhaust pipe. In fact, 86 percent of every drop does *nothing* to move the car or the driver; it is pure waste, as figure 4.1 illustrates.[13]

Energy is lost to idling, to friction in the transmission system, and to generate auxiliary power. Finally, some of the energy from the petrol starts to turn the wheels. But then there is rolling resistance in the tires. There is air that must be pushed aside. More energy must overcome the vehicle inertia of accelerating 1.5 to 2 tons of steel, rubber, and glass up to speed. That's at least 12 times as much deadweight as the average weight of the people in the car. And finally, we're cruising. Hurray! But less than 1 percent of each tank of gasoline is used to move your body around—which is what we really want. Our dream is not to drag all that steel around. We want to get to work, to a rock concert, or to visit grandma, who doesn't really fancy serving coffee to a car. She, like the boss at work, just wants the human to be there on time. That's what efficient mobility services are about.

But wasted fuel is not the only thing preventing cars from being good at efficient mobility. Ninety-two percent of the time they're parked, doing nothing. Drivers often get stuck in congestion (1.5 percent of the time) or we're looking for parking (another 1.5 percent). We drive them productively only 5 percent of the time.[14] Hence cars are very expensive products that provide no utility to us 95 percent of the time. And when we use them, they waste 99 percent of the gas we pour into them. The fuel does all kinds of silly things, like heating and polluting the air rather than getting us where we want to go. Cars even take their revenge on us by maiming, killing, and injuring us, which costs insanely much. And, by the way, they claim up to 50 percent of all city land for things like roads, parking, repair shops, gas stations, driveways, and signs.[15] Some of the most expensive, central, most attractive real estate space around is reserved for cars, not humans.

To say that such a personal mobility system "can be improved upon" is an enormous understatement. Road transport uses almost half of all oil in the world, translating to expensive imports for most countries, lowering their GDP.[16] It also causes more than 10 percent of all greenhouse gas emissions.[17] And big cars, SUVs, were the second largest contributor to the increase in global carbon emissions from 2010 to 2018.[18]

Luckily, all the solutions needed for a full disruption of this waste are available to us today: replacing cars with smart, multimodal, open transport systems, accessed by your smartphone.[19] In other words, we can jump into shared cars (autonomous or not), hail a ride with Lyft or taxis, ride e-scooters, and take electric buses, trains, and subways. We can also redesign cities to be more walkable and bikeable (including e-bikeable). Personal mobility is rapidly becoming an issue of delivering on-time services running on electricity, not about buying, owning, and parking a private car running on fossil fuels. This shift in approach and business models will profitably kill a lot of the massive energy waste from the one-person-one-car system that erupted in the mid-twentieth century.[20]

Eliminating Nutrient Waste in Food

Food is another sector where waste is ubiquitous, as mentioned in chapter 2. Once again we see energy wasted all through the system of food production and distribution—in the form of energy-rich biomass, or in nourishing, vital nutrients such as nitrogen and phosphorous, or in fossil fuels. One very clear system-wide analysis of animal products showed how 93 percent of all calories harvested from the fields are lost in the conversion when fed to animals for meat, dairy, or eggs. For conventionally raised beef, it's even worse: 97 percent of all calories are lost, with only 3 percent left in the beef itself for you to eat.[21]

If you look at the larger picture of the entire global food system's flows of dry biomass, in tons—not calories—then the same picture emerges: The total loss of dry biomass from field to table is also a staggering 93 percent.[22] But wait, there's more. A key nutrient inside the dry biomass is bioactive nitrogen. Without it, plants can't grow. After harvesting, a lot of the crop residues go back into the soil, but not enough; so conventional commercial farms spread extra nitrogen fertilizer on their fields. And a lot of it. But hang on: Only 30 percent of that extra nitrogen spread on the field is actually taken up by the roots of the plant. Which means that 70 percent of the new nitrogen applied is wasted as it evaporates as a potent greenhouse gas or is washed away, polluting waterways and eutrophicating the seas instead of nourishing the plants that should later nourish us.

And where does the nitrogen fertilizer come from? Yes, indeed, from fossil fuels, particularly natural gas extracted from deep wells, or from 100 miles offshore, or from fracking. Lots of natural gas is burned or lost due to leakage during production and transport. Then it's pumped to the refinery, where more is lost in conversion, as well as causing huge CO_2 emissions. In the end, according to one study, only 5 percent of nitrogen fertilizer's nutrients end up nourishing people.[23] The rest of the nitrogen—95 percent—is either not taken up by crops or lost in an inedible part of crop, in wasted crops, or not absorbed by the human body. The wasted fertilizer causes soil degradation, global heating, eutrophication, and overloading of natural nutrient cycles.

Finally, depending on what we eat, food can be wasted in our bodies as well. Fifty percent of Europeans are either overweight (30 percent) or obese (22 percent).[24] In the United States, 71 percent of the population is either overweight (31 percent) or obese (40 percent).[25] All this nutrient waste, fuel waste, pollution, and bad health, and we have not even counted the extensive soil loss caused by this manner of food production yet. Changing to a planetary health diet, simply by eating more vegetables, fruits, legumes, and nuts farmed in thoughtful ways, could be enough to both improve people's health and play a vital role in keeping global warming within relatively safe levels.[26]

THE BIRTH OF OUR WASTEFUL, *UN*ECONOMIC GRAY GROWTH MODEL

I could have gone through the textile and fashion industry. Or sports apparel. But it's enough. The picture is clear. How did we industrialized humans get ourselves into this mess? A large part of the answer lies in the type of economic thinking that emerged in the 1800s and early 1900s, as the industrial age took hold. That mental model (or straitjacket) sees growth as a linear flow of take-make-sell-waste. This resource-intensive gray growth model did in fact make sense in the narrow economics view of the time, when humanity was tiny in comparison to a seemingly boundless Earth with its vast forests, huge herds of wild animals, and plenty of whales, minerals, coal seams, and easy, cheap oil for the taking. In the beginning.

But by the second half of the 1900s, something had radically changed. The size of the human population, the resource use of the rich, the wastefulness of our value chains, and the ecological footprint of each person had grown far beyond the size—however vast—of just one Earth.[27] Suddenly we—meaning humanity as a whole—were using more than our planet could regenerate over time. And if everyone were to adopt the consumption of the richest nations supplied with today's waste chains, then we would need the biocapacity of around three to five planets. Nearly all people are now living in nations that consume more than their biocapacity.[28] The "magnificent" economic growth of the 1900s was—according to the mainstream economist Paul Samuelson—that century's hallmark. But the most critical characteristic of the 2000s, according to Mathis Wackernagel of the Global Footprint Network, is this planetary overshoot. We're in a new situation, a new "state of the world."[29] And the magnificent ecosystems that have supported human life on Earth for the last relatively stable 10,000 years, since the last ice age, are now in peril.

Economist Robert M. Solow is famous for formulating the dominant growth model of the twentieth-century economy. His classic article from 1956, which won him a Nobel Prize, is concise. In it he states that economic growth is driven by three main factors: labor, capital, and technological progress.[30]

The theory became standard fare in conventional economics. But anyone carefully reading Solow's article (rarely done by economists today, as economic history is not taught to much of an extent) can actually find that he consciously dismisses nature, environment, and resources as factors. In the Solowian view, they are irrelevant to economic growth.

What made his contribution new in the 1950s was not just the elegant reformulation of growth theory with a set of succinct formulas, but also that he threw out the formulation from the classical economists going back to Adam Smith and David Ricardo that growth was a function of labor, capital, and *land*. Solow held that technical progress, not land, was the "magic sauce" that made labor and capital accelerate growth. Members of the econ tribe often refer to this amplification brought on by technology as "total factor productivity," or the "Solow residual."

To keep his theory nice and clean, Solow had to throw out "land" (what we today call resources, or simply nature). Solow wrote: "This amounts to assuming that there is no scarce non-augmentable resource like land. Constant returns to scale seems the natural assumption to make in a theory of growth. The scarce-land case would lead to decreasing returns to scale in capital and labor."[31]

In a rarely noticed footnote, Solow added: "One can imagine the [growth] theory as applying *as long as arable land can be hacked out of the wilderness* at essentially constant cost" (emphasis added).[32] Solow's gray growth model means endlessly ravaging what he saw as the boundless wildernesses at no extra cost. This assumption is quite stunning to read more than half a century later. In asserting it, Solow contributed to making generations of professional economists blind to both material wastefulness and the state of the natural world. All they could see were *prices*: of labor, capital, resources, and taxes. They assumed the physical wilderness was irrelevant to civilization, as its price was zero.[33] They ignored the effects of what contemporary natural scientists have dubbed the Great Acceleration in human activity and its impacts on our natural world (see figure 4.2a).[34] And coincidentally, that acceleration started at the very same time that Solow wrote those words in the 1950s.[35]

We are just starting to pay the costs of Solow's glaring omission. Today the economy has outgrown its natural supports. There is now no more "arable land" to be hacked out of the wilderness at constant costs, as Solow assumed. *Without being accounted for* in physical units, fossil energy and natural resources have been overused as a factor of production.[36] In these first decades of the twenty-first century, we are starting to understand the full costs of this omission.[37] And the longer public economists and politicians educated in the 1900s continue to apply the gray growth model of Solow's century, the larger the losses will get. This type of gray growth is getting bigger and bigger, slowly killing off its host, like a lump of cancer metastasizing.

Today's megatrends are very clear. All analysts agree that population grows. Global wealth grows. Energy use grows. "Everything" accelerates, as shown in Will Steffen's Great Acceleration (figures 4.2a and b). Key ecological economists, such as Herman Daly, point out that if we don't head in a new direction, these megatrends will increasingly result in uneconomic growth.[38] This means that we net more bads than goods for each new unit of production the market provides. Rather than creating more "wealth"—a term whose origins lie in *well*, as in overall wellness—many countries are creating more "illth."

Let's assume car sales rise in a city already suffering from traffic congestion. More cars are sold, bringing more cash into the economy, but the city

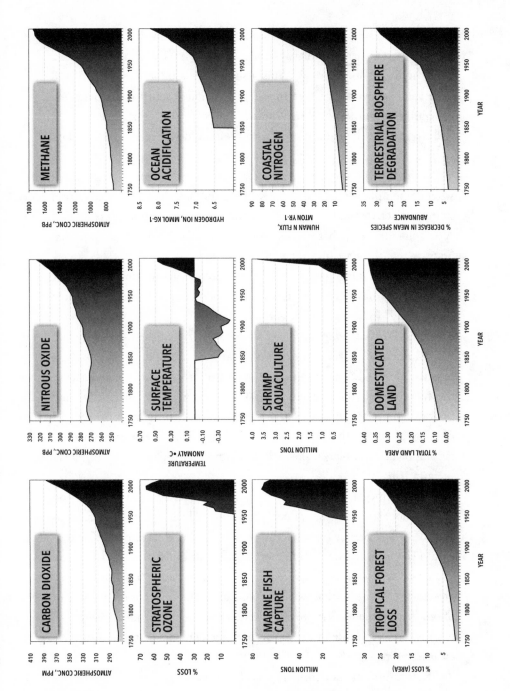

Figure 4.2a

The Great Acceleration since 1950. Earth system trends 1750–2015.

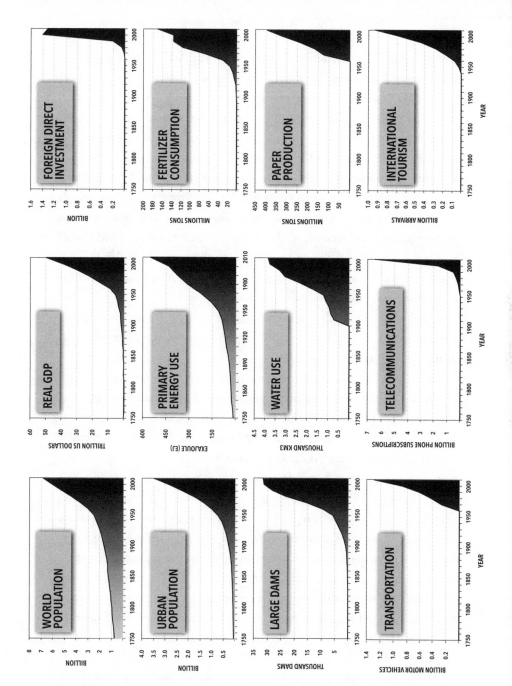

Figure 4.2b

The Great Acceleration since 1950. Socioeconomic trends 1750–2015. *Source*: Reprinted by permission of Monthly Review magazine ©. Charts created by R. Jamil Jonna, in *Monthly Review* 67, no. 4 (September 2015), https://doi.org/10.1177/2053019614564785, based on data in Will Steffen et al., "The Trajectory of the Anthropocene: The Great Acceleration," *Anthropocene Review* 2, no. 1 (April 2015): 81–98.

is growing in illth, not wealth.[39] In other words, GDP might go up, but it is an overall uneconomic growth. Each extra car adds to congestion, air pollution, and global warming while everyone driving is slowed to a crawl. The total costs in wasted time, fuel, health, and capital rapidly become larger than the economic gains of a growth in car sales. More cars go from being a good to being a bad, simply from their sheer numbers in one place. Uneconomic growth can also be seen in the rising costs of conventional agriculture, as originally fertile topsoil is depleted by excess tilling and the use of fossil fertilizer. Each harvest really costs more than it yields when all costs are taken into account. Other examples are less tangible but equally impactful: just look at how an escalation in the sale of complex financial products to customers who didn't understand them launched the incredibly expensive Great Recession at the end of 2007.

Uneconomic growth is the dark side of growth. It's also what generally results if one continues to overuse underpriced resources when the economy has become large relative to the frames and ecosystems it lives inside. Which brings us back to waste. At the heart of the issue, says Martin Stuchtey, founder of consultancy SystemIQ and previously head of McKinsey's Center for Business and Environment, is that our centuries-long free riding of natural resources has made the economy highly wasteful in how it uses energy, materials, and products.[40] Now is the time to move from waste chains to value cycles.

In contrast to uneconomic gray growth that free rides on cheap resources, healthy growth repairs this wastefulness with new innovations and practical systems solutions that can scale up quickly. It involves drastically reducing resource waste to bring down the economy's footprint as it continues to increase services and value creation in monetary terms. Conducting business this way will represent a drastic break with the latter half of the 1900s because, as figure 4.3 shows,[41] for a long, long time there has been a strong connection and correlation between economic growth and resource-use growth.

From this graph, we can see with the naked eye what history tells us: The resource use of the global economy (measured in tons) grew in an almost one-to-one relationship with the global GDP (measured in

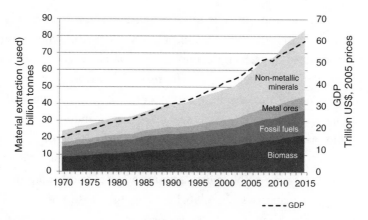

Figure 4.3

Gray growth laid bare: Global material extraction grew in tandem with global GDP from 1970 to 2015.

dollars). This is what analysts mean when they refer to a strong resource coupling of money and tons. When one goes up, so does the other. There was hardly any decoupling of the two over the twentieth century.[42] We have all grown up with this type of gray economic growth all around us. It is, unfortunately, our old "normal." As mentioned in chapter 3, this linear take-make-sell-waste economy squanders more than 90 percent of materials extracted each year after zero or just one-time use. Just as with climate facts, it may have registered in people's minds, but voters feel it is a distant, invisible problem, and are therefore unwilling to demand a shift in taxation from labor to resource use. That would have given us more jobs with less resource waste.

UNDERSTANDING BIOCAPACITY

Biocapacity refers to resources that Earth can regenerate each year, mainly from photosynthesis in plants and algae. But Earth's biocapacity is currently overwhelmed by the physical resource use. All that extraction and pollution takes a heavy toll on Earth's ecosystem, eating away at its capacity to regenerate forests, animal populations, insects, clean water, and air, and to absorb all the toxic wastes.

It is worth diving into our *ecological footprint* here. The concept is familiar, at least in vague terms, to most: a way to measure the ecological services consumed by each city, country, or person and that consumption's impact on nature. Calculating our footprint allows us to expand our accounting beyond just the immaterial money flows that conventional accounting relies upon. It also allows us to "follow the matter" and see Earth's biocapacity more fully—to see its ability to support our needs more clearly.

As the basic unit of biological accounting, ecological footprint and biocapacity calculations use a global hectare (gha). This represents an average area of land and sea that can produce and regenerate what sustains us: where plants can grow, absorb CO_2 by photosynthesis to make biomass and oxygen, and filtrate water, or where crops can thrive, or wildlife can find habitat if enough natural biocapacity is left unused by humans. The ecological footprint includes the freshwater that supplies our homes and industry, and the ocean habitat that supports our fisheries. It also represents sinks—on land or in water—that can absorb our excess carbon and other waste. The renowned ecologist E. O. Wilson has suggested that humans should leave half of Earth's biocapacity for other species, to keep about 85 percent of biodiversity intact.[43] This measure of natural capital has attracted much attention, enthusiastic and critical.[44]

A group's ecological footprint is not static; it can change with consumption patterns and technology. It's no surprise that footprints per person in rich nations far exceed those in poor ones. The Earth's total biocapacity, though, doesn't change much. As Mark Twain once observed, they don't make much land anymore. And the rainwater we get annually is all the freshwater we'll get from the skies. We can certainly restore water cycles to ensure more water is accessible to us, to other animals, and to plants, and we can regenerate land to make it more productive in an ecologically sensitive way. Yet, even if we achieved such improvements, we are far overshooting Earth's biocapacity with our current levels of waste.

Our blue, white, and green ball floating in space has twelve billion global hectares in total. Since we're now around 7.8 billion humans in 2020, that means there are 1.5 global hectares per person, equal to 3.7

acres of average biocapacity per person. Humanity uses more than that, however. In 2018, we used 2.5 gha per person, or 6.2 acres per person. That means humanity is now using the biocapacity of more than 1.7 Earths.[45] And that is on track to double again by 2050. Hence, if humans continue business as usual, we'll need the biocapacity of three Earths to supply and deal with the environmental impact of our excessive resource use. About 60 percent of the human footprint is currently energy and carbon related.[46] That means most of our footprint results from the demands that global greenhouse gas pollution puts on nature's sinks to store safely.

BACK TO THE PARADOX

And so we wind back to the growth paradox that we pondered in the introduction: We must have more growth, yet more of the same type of growth is impossible as we're running out of Earths. To solve this paradox, we must be very clear about what we mean—conceptually as well as mathematically—when we talk about healthy growth.

The basic idea is that in the coming decades GDP can still grow (at moderate rates in rich countries, higher in poor countries), while our environmental footprint and social inequality move sufficiently rapidly into acceptable bounds. By acceptable, I mean a reasonably safe risk for current and future generations of both humans and other species. The value creation from services and goods (measured in money) can continue to grow while wastefulness (in tons and gha) declines at a rate of change sufficient for the total to soon fit well within the biocapacity of one planet. It's the art of changing year-by-year, from now to 2050 at the latest, toward one-planet-compatible, equitable living.

Green growth refers to the footprint part of that equation: the type of economic growth born from wasting fewer resources year by year, resulting in a measurably smaller environmental footprint from all economic activities. *Healthy* growth refers to green growth in which the economic benefits are sufficiently inclusive of all humans. In healthy growth, human well-being keeps rising while the footprint falls and social trust grows.

Or to put even more simply: Healthy growth solves the paradox by turning conventional economic growth around in a greener and fairer direction: Healthy = profitable × green × just.

There will be pitfalls along the way, and points at which it's hard to distinguish gray growth from green growth and healthy growth from both. Are there clear and simple metrics that will help with such navigation? Chapter 5 provides a compass.

5 TELLING GRAY GROWTH FROM GREEN GROWTH, AND GREEN GROWTH FROM HEALTHY GROWTH

We can no longer afford the gray growth model, nor all the greenwashing of gray growth that has partly discredited the idea of green growth. In a world that is using more than 1.7 planets of biocapacity, any more gray growth fails to achieve what needs to be our number-one goal: make the economy thrive inside our one Earth again by cleaning up the wastefulness from the twentieth century. This overhaul must start immediately and be completed no later than 2050 in order to minimize risks of triggering irreversible decline in Earth's life-supporting systems.

The idea behind green growth is to describe a cleaner pathway from today's economy that eats up the Earth to a healthy economy that can flourish safely within Earth's boundaries for centuries. It combines two main dimensions: the change of value creation (economic production in dollars) from year to year, and the change in environmental footprint (the impacts of materials use in tons) from year to year. Hence, green growth means more economic value created with less new physical material extracted, while reducing the environmental impact of those material flows. We get a gradual dematerialization of the economy—or, as an accountant might say, of each dollar of "value added."[1]

Today, what grows much faster than in the gray growth of the 1900s is the value created from near-weightless services and experiences—from entire digital realms, the smart grid, education, human services, and healthcare to cultural experiences, entertainment, and consulting services. The small smartphone, for instance, has apps that have replaced hundreds of other products and separate devices, such as cameras, CDs, CD players, books, maps, videocassettes, alarm clocks, voice recorder, tickets, catalogs, photo albums, etc. One smartphone can cut resource use by 91–99 percent compared with buying all that.[2] And the physical materials still embedded in these new lightweight, digital parts of the economy can increasingly be reduced, reused, recycled, and upcycled, as with Apple's disassembly robot Daisy.[3] If well configured, 3-D printing can increase efficiency, logistics, reuse, durability, and maintenance of a wide range of products.

For real green growth, this dematerialization needs to happen overall at a faster annual pace than the GDP growth in each country. In short, green growth makes *more value with less waste*. It gets radically more value creation out of any resource we use, so that our physical footprint on Earth's biocapacity and boundaries can decline every year. The good news is that it has already started in many countries around the world. This type of growth can and should be improved and accelerated continuously, until nature again has abundant biocapacity reserves to regenerate itself every year. Only then can the footprint stay in a steady state (or decline) while GDP continues to rise to solve poverty and inequality. The unused biocapacity that humanity no longer grabs can then regenerate and evolve nature's amazing beauty and healthy biodiversity. Good for us humans, and for all the other life around us, from fungi in the soils to whales in the oceans, amphibians in the wetlands, beavers in waterways, and beetles and bears in the forests.

THE GREEN GROWTH COMPASS

We thus arrive at two main dimensions of growth: one that describes an increase or decrease in economic value creation (gross profit at corporate level, GDP at national level) and another that describes an increase or decrease in the environmental footprint (EF).

In the following, I'll speak of three related measures of environmental footprint: carbon footprint (the greenhouse gas emissions in tons), material footprint (measured as material consumption of biomass, fossils, metals, and minerals added together in tons), and ecological footprint (in global hectares of biocapacity). All three are methods to measure the environmental footprint of any economic activity. Depending on usefulness for the issue in question, data availability, and quality, any of these may be best suited when total resource use and resource productivity are discussed, as when defining green growth below. Whichever we use, they tend to be highly correlated.[4] Rather than waiting for the *one* perfect metric, companies and countries can focus on continuous improvement right away, fine-tuning the accounting along the way.

This gives four possible models that any country or corporation can move along over time. I call it the *green growth compass*, and anyone can use it to navigate the economic growth landscape.[5] Figure 5.1 shows these four different models of economic growth.

If there is economic growth (GDP up), but the growth uses more physical resources, increasing its environmental footprint (EF goes right)

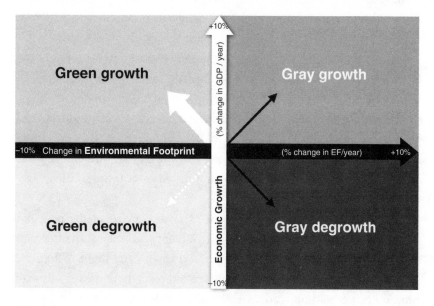

Figure 5.1

The growth compass shows the four possible directions of growth, based on the change in value added (GDP) and the change in environmental footprint (EF) per year. EF can be measured in tons of greenhouse gas emissions, materials consumed, or global hectares of biocapacity, depending on data relevance and availability. The bold white line shows the improvement of resource productivity (more value created with less footprint).

every year, then we're in *gray growth* territory. That's the northeast (upper right) quadrant—the typical realm for 1900s economic growth, and still the dominant realm for the majority of countries in the 2010s.[6] In the gray growth model, a country's economy typically makes more market value for its own people while using so much more oil, soil, trees, fish, water, and other resources every year that nature is left in a worse situation. Extreme examples of gray growth are easy to find: the United Kingdom in the 1800s, the United States in the 1900s, and China in the early 2000s. In gray growth, there are often efficiency gains (more output per resource unit). But the volume of the growth eats up the efficiency gains. The result is a higher footprint too. This catch-22 is known as the rebound effect: Efficiency brings down costs and resource use in individual products, but those lower costs encourage more consumption of other products, which begets more resource use overall. It's part of the growth paradox—specifically the Jevons paradox, named for William Jevons, who in the nineteenth century observed that transforming coal into energy in more efficient and thus cheaper ways led to people using more, not less, of it.

This paradox underlies one of the biggest green-growth worries of the environmentally alarmed: what may initially look like promising green growth may backfire and end up pushing *gray* growth.

There are three main kinds of rebound effects—direct, indirect, and economy-wide. If the auto industry makes cars more fuel efficient but the fuel savings encourage more people to drive longer, that would be a direct rebound. The magnitude of the rebound depends on people's responsiveness to the resource price.[7]

An indirect rebound happens when, say, the owner of the new and better SUV doesn't really drive more than before but spends the fuel savings on some other type of consumption, like an extra flight to Florida. Even something as green as a windmill has indirect rebound effects. It may produce power without carbon emissions, but it takes embodied energy, steel, and concrete to make and install it, and other resources to maintain it, operate it, and distribute its energy.

An economy-wide rebound represents the sum of all direct and indirect effects from better efficiency rippling through the entire system. This is what

Jevons observed in the United Kingdom during the 1800s. New, improved steam engines needed less coal to do the same work. But the improvements resulted in seemingly ever-growing coal use as the new engines' popularity skyrocketed. The relative cost of using a resource decreased, which increased the quantity demanded. The Jevons paradox suggests that more efficient (and hence comparatively cheaper) energy leads to faster overall economic growth, which again increases resource use throughout the economy.[8] Some see rebound as inevitable for all resource productivity solutions. Its effects, they say, make efficiency savings impossible.[9]

So, will all efficiency improvements inevitably rebound? Is resource productivity indeed futile as a solution to resource overuse? A lot of research—often complex and with varying definitions—has tried to figure this out since around 1980. But findings from the last decade are quite clear.

Direct rebounds—when they happen—tend to be rather small. Rebound rates varied depending on a number of factors, but studies typically show that in richer countries, with generally high levels of overall consumption, only 0 to 30 percent of resource savings is lost to rebound.[10] Many early studies that showed higher rebound used twentieth-century data collected long before radical efficiency or green growth concepts got the faintest chance to influence real policies. And also, long before cheap renewables became a reality, digital end-user efficiency and circular economy practices gained traction in the marketplace. All these factors can reduce rebounds and create a greener growth.

Indirect rebound effects can be somewhat larger, often 20 to 40 percent of the savings, if good design and material flow analysis of the value chain is not taken into account.[11] But the rebound can be reduced by shifting to renewable energy down the supply chain, shifting from waste chains to value cycles, and greening procurement in all companies as well as governments. Consumer savings that might feed more consumption can even be redirected into further efficiency investments, such as saving for an electric car rather than the extra trip to Florida. Some studies show that this kind of efficiency with increasing returns makes even super-conservation possible.[12]

Certainly, rebound effects can undermine piecemeal efficiency efforts and need to be taken into account when planning. Yet, fretting about strong rebound is mainly a distraction.[13] Any remaining economy-wide rebounds are best handled with coordinated policy (see part III). For example, controlling fossil fuel use and emissions with cap-and-trade measures, increasing efficiency standards, or imposing green taxes on resource use all help to socialize the gains from efficiency innovations. Without such measures, the shift from gray to green will likely be slower.

When GDP growth happens with *less* resource use from nature than the previous year, then we are moving toward *green growth* in the northwest (upper left) quadrant. When a country manages to move into and stay inside that quadrant year for year, it means that the economy is making more value for people while leaving nature in relatively better shape each year too. Decreasing footprint can be done by emitting and extracting less while reinvesting in nature. These waste-killing and regenerative practices are what moving in the direction of better resource productivity is all about (the bold white arrow in figure 5.1). Is this truly possible? Yes. Sweden is one example of a country that has improved its carbon footprint every year since 2000. Denmark, the United Kingdom, the United States, and fourteen more countries are also heading in this direction.[14]

If GDP declines from one year to another, then this reduced production and consumption may reduce the physical resource use, too. This means a lower environmental footprint, with less pressure on rivers, forests, soils, and oceans. When both the GDP and EF go down each year, then we're in a *green degrowth* (the thin dotted white arrow in figure 5.1). The challenge with degrowth is that as sales go down, companies start to fire employees, probably increasing unemployment and inequality. And private investments in new (risky) opportunities go down, slowing the speed of structural change in the economy. This happened to Russia in the 1990s, after the Soviet Union collapsed: Many of its dirty state industries were stalled when the subsidies stopped, and resource use, as well as CO_2 emissions, dropped accordingly. Japan from the 1990s to 2010 is another example: the economy contracted during what was dubbed the "lost decades," and emissions fell somewhat, too. During both the 2008 financial and 2020 corona crises, the United States and the European Union were also in this quadrant: GDP shrank, as did emissions.

The fourth quadrant (southeast, or lower right) is one that nobody wants to inhabit. In *dirty degrowth*, GDP falls while the EF increases, as happened in Venezuela since 2013. The economy crumbled, and the rich emigrated. Everyone became worse off, particularly the poor and vulnerable. Today, dirty petroleum production chugs along with little or no investment in cleaner production. Some other industries, such as gold mining, have seen an uptick, but there is no money to clean up the mining waste, which then poisons the rivers. When gray degrowth takes hold, there can be no money to enforce compliance. Things can become increasingly corrupt, downright illegal, and violent. The hungry start to chop down trees around them to cook, start illegal wildlife hunting, or to steal food from restaurants or local farmers. Heading in the dirty degrowth direction for a while soon makes human lives more "nasty, brutish and short," in the words of Thomas Hobbes. Nature, too, goes down the drain. Japan has also been visiting this quadrant, particularly in the years following the 2011 Fukushima nuclear accident: less GDP and more pollution.[15] Syria since 2011 is a third such case.

A couple of things are key to remember. First, growth always means change over time. It says nothing about the absolute level of anything in the economy. The growth compass is thus a compass for orientation: In what *direction* are we moving over any time period? Second, resting our future goals on GDP alone has many critics, and nearly all professional economists agree with those critics at least on one count—that GDP alone is *neither* a good goal nor a measure for well-being development.[16] GDP is simply the total volume of all economic activity in any year, based on market prices and embracing the good, the bad, and the ugly. Sure, GDP can rise with the proliferation of good things like solar energy or wind power. But bad things can drive up GDP, too. Huge wildfires, for instance, generate a flurry of economic activity. Large brigades of firefighters are hired. Homes are burned down and must be rebuilt. Lots of people need medical care. Similarly, if a terrorist throws a bomb into a mall, then the hunt, the cleanup, and rebuilding will all increase GDP. Wars, too, can fuel economies in insidious ways.

These are certainly not the kind of GDP increases that anyone really wants. Even so, I'm personally not a big fan of GDP.[17] It measures economies imperfectly and with large omissions and margins of error.[18] One can only *hope* that, on average, any growth in GDP per person contributes to improving human well-being. It tends to do so in poor countries, but in already wealthy countries, that contribution is weak or nonexistent.

Still, GDP is one of the best-known, most widely used economic metrics we've got, and something that "everyone" relates to. That's why, if we're careful about how we use it, GDP can be put to good use as an indicator of the *activity level* of the entire market economy with a relatively well-documented, long-term series of data available to us. Because it is, after all, these activities (the driving, logging, farming, mining, fishing, flying, manufacturing, computing, video streaming, healthcare, service delivery, and so on) that cause the total resource use. And it's the activity level that tends to generate jobs—or, if it falls, unemployment—for people. There have been many attempts to modify GDP or replace it with something else—like indicators of genuine progress, or sustainable development, or even happiness. But rather than wasting our time on GDP bashing or somehow struggling to correct it for its many faults, we can integrate existing GDP numbers with additional measures to tell whether any observable change in future GDP is the type of growth we want—or the opposite.

GDP is a little like a tachometer, the device that in combustion engine cars monitors the revolutions per minute (RPM). It doesn't show where we're going or if we're going uphill or downhill. Neither does it show if we're speeding toward our vision or moving like a snail. It's useful to show whether the economic engine is revving too low or too high. Along with many others, I believe we can supplement GDP with a broader set of a few meaningful key metrics that will give us a fuller economic dashboard. We can then reframe it into something useful, rather than continue to fight it or worship it. We can even be agnostic about GDP, neither for nor against it. Because there is no reason to expect GDP numbers themselves to give any account of human well-being, trust or natural capital.[19]

One approach to a meaningful use of GDP is to see it in connection to fossil-fuel use and greenhouse gas emissions. We then get an indicator of the economy's "carbon productivity": How many dollars of value creation do we get for each ton of greenhouse gas emitted ($/tCO$_{2e}$)? Now, that's an interesting metric for anyone interested in systems change to solve global heating.

Some directions of GDP growth are beneficial for society over time. Moving into the upper left green growth quadrant reduces the pressure on nature measured in tons of resources used while making people better off in terms of purchasing power. It also potentially benefits the poor (the topic of chapter 6). If we get genuine green growth right, we can grow companies, cities, industry sectors, and the entire economy in a direction that brings us to one-planet-compatible living. Humanity can keep flourishing, economically and otherwise, inside the safe boundaries of the planet. Defined by the typically feel-good words *growth* and *green*, it has the air of a win-win game. That's why phrases such as "green growth" and "green new deal" have been used frequently by major national and international players since around 2008.[20]

The usefulness of the growth compass lies in cleaning up the fog of confusion around these words and concepts: Many politicians, executives and international agencies acknowledge that conventional growth is environmentally unsustainable, or gray. Most people recognize that material resource use is already too high. It's causing global warming (from fossil fuels) and vast losses of natural, beautiful wildlife (biomass). It's filling the oceans with toxics and plastics (one-time consumption). And these trends seem set to continue to rise indefinitely. Hence the need to talk of greening future growth. But *speaking* about green growth does not change the environmental footprint. Only real-world investment and thoughtful action, by both companies and governments, both cities and citizens, can do that. The all-too-human response has so far been to chatter away about sustainability and green growth while continuing to walk in the gray growth direction. But this is *greenwashing*: pretending to be walking in another direction than you are. Speaking about northwest while walking northeast is not helpful. You'll get lost but pretend everything is just fine. This is where psychology come to the fore—when double speak and self-justification move in to cover over the dissonance.

Some improvements are often made even while wandering astray; the gray growth is made a little less gray. Corporations that are clever at greenwashing highlight a few green things they do. They might plant some trees, or put ecological soaps and efficient LED lighting in the bathrooms of oil rigs drilling at full speed. Or they might aim for something more directly related to their products—like a food company that tightens its certified palm-oil sourcing enough to tout its efforts but not sufficiently to end its complicity in vast fires burning in tropical forests to make way for palm plantations. These small strides are—of course—far from sufficient in that the total ecological footprint still grows in physical, absolute terms, even if a tiny bit slower than previously. A little less gray is far from genuinely green.

Yet for our work, and in this book, real green growth is considered not just wishful thinking or a theoretical possibility. There are success stories of eliminating wastefulness at company, city, national, and regional levels as we'll see in chapter 7. Genuine green growth is a new growth model that is very different from twentieth-century gray growth. It only really got going during in leading countries and companies during the 2010s and is now delivering the first steps toward a healthier economy. When accelerating green growth year after year, and rapidly scaling it to a global level, we will be able break the strong historical coupling between global resource use and global GDP. The economy will then gradually change from its current high-risk track of destroying ecosystems to one that thrives within them.

The question is whether we're seeing quick enough green growth to change the system in time. And what exactly does that new, fully changed system—a green economy—look like? The short answer is this: A thriving green economy will mean about double the volume of global GDP from 2020 to 2050, with less than half the ecological footprint, while human greenhouse gas emissions are approaching net zero.

DEFINING AND MEASURING GREEN AND GRAY GROWTH

To precisely track the speed of system change, we need some very clear definitions and simple equations. It's time to become somewhat more technical. First let me revisit the confusing mix of green growth definitions from chapter 1: International bodies like the World Bank, European

Commission, the United Nations Environment Programme (UNEP), and the Organisation for Economic Co-operation and Development (OECD) have all proposed complex definitions but with no clear, quantified criteria in them. OECD's, for instance, is: "Green Growth means fostering economic growth and development, while ensuring that natural assets continue to provide the resources and environmental services on which our well-being relies."[21] In collaboration with professors Jorgen Randers and Johan Rockström, I have developed a much simpler and more communicable one that can be practical for both companies and countries in a verifiable way. Our formal definition is: *Green growth is an increase in economic output that lowers total environmental footprint.*[22]

"Economic output" is best understood as the value created by economic activities over time, measured in dollars. Footprint is the economic activities' impact on the environment and ecology. And "total environmental footprint" can be measured in a number of ways—such as CO_{2e} emissions in tons per year, in domestic material consumption (DMC) in tons per year, or by ecological footprint measured as global hectares of biocapacity.[23] In principle, we can include any material resource use that directly relates to ecosystem and planetary boundary dimensions. It is particularly important to include the four boundaries (climate, nutrient, land use, and biodiversity) that have already been pushed outside the safe operating space for humanity.[24]

The above simple definition can then be used to define green growth with precision. Here are the core concepts and their shorthand notations as this book uses them:

- VA (value added) = the value creation in economic activities = the monetary value of all newly generated goods and services less the value of all goods and services consumed in their creation, in $.
- GDP (gross domestic product) = a nation's total economic activity = the value added of all final goods and services produced within a nation's geographic borders at market prices, usually for a year, in $.
- real GDP = inflation-adjusted GDP, i.e., in constant prices (as in 2011 $).

- g = rate of change in real GDP, i.e., economic growth, in percent per year. (This is what everyday speech often calls "growth was 2 percent.")
- gha = global hectares of average biocapacity (from photosynthesis).
- EF (environmental footprint) = the environmental impact of economic activities, in tons/yr or gha/yr.
- ef = rate of change in EF = the rate of change in environmental footprint, in % per year.
- RP = resource productivity of value creation = real GDP/EF usually for a year (in \$/gha or \$/tCO$_{2e}$).
- rp = rate of change in RP (resource productivity), i.e., resource-productivity growth, in % per year.

Note that capital letter initialisms refer to real-world measurement variables (such as \$, tons, or \$/ton), while lowercase letters refer to the *rate of change* in these variables. As a variation on Thomas Picketty's famous $r > g$[25] we can now define green growth in a very simple yet crystal clear way, with only four characters:

Green growth is: $rp > g$

Which can be read as "resource productivity growth is greater than economic growth." In other words: growth is green—or at least heading in a green*er* direction—when the rate of change in resource productivity is greater than the rate of change in GDP.[26] It means more immaterial money with ever less material wastefulness.

Thus, any company, city, sector, or country can only have green growth when resource wastefulness is conquered faster than the growth itself. We get even more efficient and smarter physical resource use for each dollar value added. This gives higher improvement in resource productivity (rp) than in consumption growth (g) over the same period of time. To illustrate: If Sweden sees a GDP growth of 2 percent per year, and at the same time uses fossil energy so much smarter that the carbon productivity improves by 5 percent, then the country displays green growth in the climate dimension. The economy grows larger in real inflation-adjusted terms, yes, while at the same time emitting ~3 percent less annual greenhouse gas emissions.

Moving in the direction of green growth, then, demands absolute decoupling of GDP growth from resource use. The economy grows toward the green growth quadrant through eliminating so much wastefulness that the footprint from emissions and resource consumption falls.

But that doesn't mean that *any* rate of change is sufficient. In chapter 7 we'll discuss how quickly resource productivity needs to grow each year to be sufficient relative to planetary boundaries. In the meantime, let's take a closer look at gray growth.

Gray growth is the dirty twin of green growth. It can be defined as an *increase in economic output that also causes an increase in the total environmental footprint.* Here the resource use (and associated environmental footprint) grows in spite of a somewhat improved resource productivity. Each new car may have a somewhat more efficient combustion engine, but since more cars are produced and/or driven more, the total environmental footprint from cars still goes up. Here, we see the rebound effect in full action. In equation form, gray growth looks like this:

Gray growth is: $rp < g$

Let's assume Norway has a GDP growth of 2 percent in one year, but the country's resource productivity only improves by 1 percent. A 2 percent larger economy that uses resources 1 percent more effectively will increase its total environmental footprint by ~1 percent per year. In such gray growth, the volume of growth in economic activity eats up all the efficiency gains. The economy grows along with a (somewhat smaller) growth in emissions, despite the tiny gain in efficiency per unit. It just ain't big enough.

The whole point of these very short and simple formulas is that we can now both allow for growth and hold companies, cities, politicians, and nations accountable for achieving ecological targets, with the same criteria across the board. Do they walk the talk of eliminating wastefulness? And do they do it fast enough? Or are they only greenwashing—delivering a few improvements here and there, but overall continuing the same old type of gray growth while speaking about green or clean or sustainable growth? It can also help us set meaningful targets and rankings (which I'll return to in chapter 7).

The main reason for the historic tight coupling of resources and GDP is that resources such as coal, oil, and timber have been abundant and cheap while skilled labor has been relatively scarce and expensive. Thus, it made sense for companies the world over to invest in machines to improve labor productivity (value added per hour of work). Profits were made by running machines on cheap fossil energy (tractors and trains, excavators and elevators, cranes and computers) so that fewer and fewer workers could produce ever more. This has been the almost singular focus of innovation since the 1800s and a powerful driver of economic growth.[27] It has given incredible results: Labor productivity has improved 200-fold in many sectors over the last two centuries. And it is still improving, with a global doubling since 1970, even if at a declining rate lately in richer countries.[28]

From figures 4.1 to 4.3 we now understand just how many resources have gone into making economic growth happen. If we continue in the same vein, the growing "dark side" is easy to imagine: More gray growth will undermine the planetary ecosystems while inequality and social tensions escalate. Social exclusion, surveillance, and wars may follow, as ecosystem, climate, and societal breakdowns reinforce each other. Three types of dark side futures will be outlined in the scenarios "Same," "Faster," and "Harder" presented in chapter 11.

Abundant gloomy scenarios cloud the horizon with imagery of an uninhabitable Earth. It's hard to imagine, argue, and describe in a credible, inspiring way how humanity may avoid such dark futures. It can seem unlikely, unreasonable, naïve. It would be truly surprising if we do, as it requires a huge systems change. This book attempts the unreasonable: to describe, measure, and guide toward the surprising pathway that may— just barely—take us through after all. A pathway on which the world may achieve (most) sustainable development goals within (most) planetary boundaries by 2050.

To break away from our current trajectory of continued gray growth requires a fundamental shift in economic logic. Since the Earth seemed vast in the early 1900s and skilled people relatively few, it made economic sense to focus on the scarcest factor: labor (just like Solow did). Now, however, we're in a world of nearly eight billion people, heading toward nine billion or possibly more, and we're running out of biocapacity, particularly of sinks to deal with the total waste. This is a new pattern of scarcity.[29] Rather than continuing to exclusively multiply labor productivity to boost economic growth, we need to shift innovation, markets, and incentives toward also *multiplying resource productivity*. To deliver, the growth compass tells us we need to shift course by 90 degrees: To rapidly head northwest, rather than continue northeast.

THE MAIN SOLUTION: *SUFFICIENT* IMPROVEMENT IN RESOURCE PRODUCTIVITY

It's crucial to understand why the shift in innovation from *labor* to *resource* productivity is a paradigm shift toward a better version of capitalism. Different growth directions yield a different type of economy over time. You can see the effect of these growth directions in the coming decades in figure 5.2.[30]

As the illustration shows, the twentieth century was characterized by gray growth. When global GDP went up, so did the ecological footprint, along with a continual rise in greenhouse gas emissions all the way up to 2020. By the 1970s, our economy had expanded beyond a one-planet footprint, mostly because of these emissions, according to Global Footprint Network.[31]

The curve labeled "Gray growth footprint" shows where we might end up if the twenty-first century also follows a gray growth course, giving us a full two hundred years of expanding human footprint. The footprint rises more slowly than GDP but still increases (by around 1 percent per year). This is called *relative decoupling*. The curve for "Green growth footprint," however, shows how, with real green growth, the total environmental footprint will go down even while global GDP continues to rise, albeit more slowly. This is called *absolute decoupling*.

On the green growth curve, the ecological footprint falls to nearly half of its 2020 level before 2050. This type of green growth results in

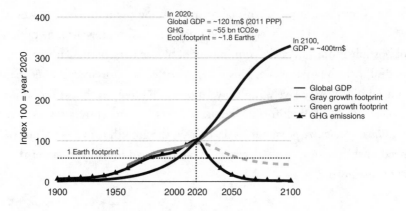

Figure 5.2

Gray vs. green growth over two centuries, relative to the year 2020, which is 100. GHG refers to all greenhouse gas emissions from human activity, showing historic data from 1900 to 2020. The gray growth footprint line shows the historic trend of the global ecological footprint from 1960 to 2020, and continued gray growth to 2100. The green growth footprint illustrates how green growth can break the close coupling between a growing GDP and its associated ecological footprint from 2020 to 2100. The figure shows real-world data up to 2020, and scenarios by author from 2020 to 2100.

more than double the value created, but with less than half of the footprint. The resource productivity (GDP/footprint) will then have increased by a factor of four (2/0.5 = 4) or more. Such "Factor 4" or even "Factor 5" and "Factor 10" examples and developments have been studied and described extensively.[32] To become widespread reality, economies around the world—starting with the rich—must decouple their footprints annually in absolute terms and at sufficient rates of change to matter.

Any economy whose GDP continues to rise due to a growing population and/or higher incomes per person will demand more goods and services. The associated resource use, emissions, and ecological footprint will tend to go up unless decoupling efforts are more pervasive than the economic growth rate. Sufficiently improved resource productivity is what makes true decoupling possible. It must be planned for, invested in, executed, and monitored year by year, just like labor productivity used to be. The sole difference between the gray and the green footprint pathways, between unstable or stable Earth systems in coming decades, lies in low or high rates of change in resource productivity.

But why put so much emphasis on resource productivity? Would it not be easier and more straightforward to simply demand cuts in emissions and legislate reduced resource use, rather than assigning all this priority to resource productivity? There are three key answers. The first lies in incumbent social structures, corporations, and vested interests. Such negative counter policies ("Cut consumption!" "Stop emissions!" "Ban fossils!" "No more extraction!") provoke way too many obstacles and political opposition. When people feel their jobs and economic freedom are threatened, they will mobilize against you with their force. As a strategy for societal change it may be better to go "with the flow"—to reframe growth as a future opportunity rather than try to force contraction, degrowth, or steady-state mandates onto growth-addicted corporations with thousands of shareholders or election-conscious politicians with millions of rich-country voters to whom they have promised growth and who sincerely believe they want it and need it. With sufficiently high rates of resource productivity growth, corporations can keep growing from an ecological point of view. A rapidly growing offshore windmill company will cause more CO_2 emissions from installing the windmills but will still increase resource productivity overall, as long as fossil energy is being outcompeted.

Second, as important green growth innovations get going in most sectors and corporations, they can outcompete the gray growth business by sheer competitive market forces. This may create a faster turnaround than betting only on bans, taxation, or regulatory routes. Tesla is one example of how its rapid growth has been crucial to incite innovative electrification of mobility in the world's entire car industry, including buses and trucks.

Third, the human brain: growth feels more metaphorically "up" and solutions focused than "down" and "no!" And resource productivity is all about fixing wastefulness in an "up" way: by turning waste to wealth. Everybody, right, left or center, reacts negatively to waste. We want more with less waste! It's a psychological growth mindset built on opportunities and new frontiers rather than on austerity and puritanism.

Since around 2000, some rich countries, including Sweden, Switzerland, Denmark, and the United Kingdom, have begun decoupling their footprints somewhat. Since 2005, at least eighteen more countries have followed suit, including the United States. Since 2010, the decoupling trends are clearer in carbon, material, and ecological footprints.[33] Yes, several societies are dematerializing their economies,[34] but not yet quickly enough. Wastefulness—relative to best available solutions—is still pervasive across all nations and in all sectors. How do we scale solutions to actually do it quickly enough?

These two simple inequalities ($rp > g$ or $rp < g$) give us the tools to discern whether any economic activity is moving in the green growth direction over time. But even if absolute levels of resource use go down, how can we know if the trajectory is green *enough* to be healthy for humans and the planet in the long run?

This brings us to the crucial issue of *sufficient* annual change in resource productivity. What are the necessary rates of change? How much higher must rp be than g to qualify as sufficient? In order to answer this, we must have some science-based targets that draw on consensus Earth system science. We need some clear, simple metrics to see if any growth in value added is sufficiently green and just. The challenge is to break such natural science and often global numbers down into metrics that make sense at the national, city, state, and corporate levels in a fair way.

In chapter 7, we'll crunch the numbers on science-based targets and offer answers to these questions. But before that we'll outline how to take green growth to a new level, one that makes it a road to a truly healthy economy.

BEYOND GREEN GROWTH: THE HEALTHY ECONOMY FRAMEWORK

In chapter 1 we saw that, of the four archetypes of growth, the "linear and more" model frames the popular understanding of GDP growth in politics and in media. Within this simplistic model, the only way to grow is upward and bigger. It's like the southwest-to-northeast diagonal in the growth compass: up means more of both money and resource use. If one falls, both fall. And falling down means fearful failing within this conventional mindset.

As we saw, there are, however, other directions and other types of growth. And there are other ways of evaluating progress and prosperity.

They mirror the growth of the human body. As children, we grow up, get taller, but then—after about eighteen years—enter into more complex patterns of growth. We also enter into more complex relationships with the people and nature around us. In addition, humans "grow" by having their own children, too, and socializing them into good citizens who have trust in and engage with society.

In the same vein, conventional economic growth must mature in the 2020s. Even green growth must mature and become more socially inclusive. The grown-up counterpart is the healthy economy, a framework that shifts our focus away from measuring wealth only by the annual flow of gross profits or GDP and toward measuring wealth based on broader capital stock—a mix of productive, natural, and social capitals. In a healthy economy, a country's future well-being relies not as much on the annual ups or downs of its production (which GDP measures) but on how well the country is using that growth to build its broad capital stock (its "productive base"). Emphasis is placed on national balanced broad *wealth*, not only higher national *incomes*. With balanced wealth, I mean that at least two capitals increase while the third does not decline. The change in a nation's wealth can be measured as the change in value of all the capitals in a balanced way over time, as outlined in the lower row in figure 5.3.

The first fundamental distinction to be made in healthy growth, then, is between yearly flows and the long-term development of stocks. In our own personal economies, we know this as the distinction between our incomes (monthly, yearly) and our wealth (the accumulated assets over years). The upper row in figure 5.3—the flows—is the flurry of economic activity happening each day, month, quarter, and year. The stocks, however, make up our stable broad asset base that make the flows possible.

Each day, workers get out of bed and go to work to create economic value. This flow of daily work comes from the human skillsets that make

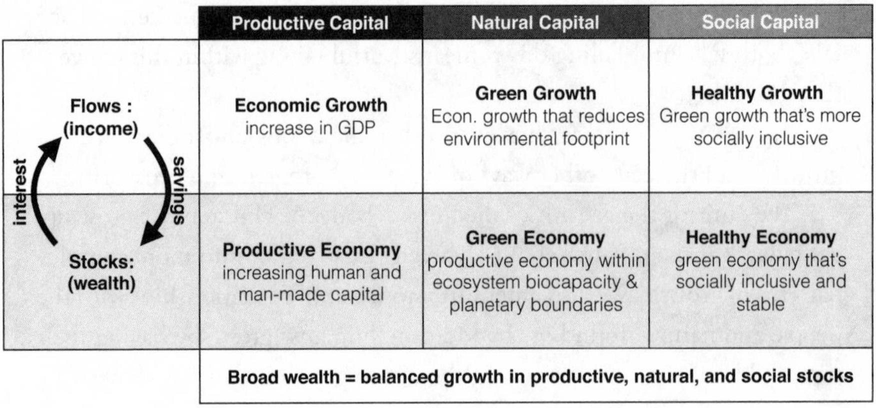

	Productive Capital	Natural Capital	Social Capital
Flows : (income)	**Economic Growth** increase in GDP	**Green Growth** Econ. growth that reduces environmental footprint	**Healthy Growth** Green growth that's more socially inclusive
Stocks: (wealth)	**Productive Economy** increasing human and man-made capital	**Green Economy** productive economy within ecosystem biocapacity & planetary boundaries	**Healthy Economy** green economy that's socially inclusive and stable

interest *savings*

Broad wealth = balanced growth in productive, natural, and social stocks

Figure 5.3

The healthy economy framework, showing annual flows (income) at the top line and long-term wealth as the broad set of assets (bottom line). Twenty-first-century growth is all about increasing the broad wealth and how savings are used to maximize all three capitals to create a healthy economy.

up the wealth of *human capital*, which comprises the largest part (about two-thirds) of the productive capital. The rest is mostly human made machines, buildings, and infrastructure. Let's illustrate *natural capital* with an acorn: The acorns are the flows; the oaks are the stock. The natural capital stocks that a green economy relies on are all the forests, fields, soils, and waters that nature uses to regenerate itself. This unused biocapacity reserve hopefully enables the productive economy to fit well within the safe zone of the Earth's planetary boundaries. In *social capital*, the day-to-day interaction of citizens and the trusting relationships they embody as they interact with each other are the *flows* of social capital. And their basic trust in stable institutions, inclusive practices, laws, and fair enforcement over time constitute the stocks of social capital.

In a productive economy, all financial capital depends on strong social capital, and both depend on a strong natural capital. If this is unclear, just think how about how a city's real estate value crumbles if under threat of war, extreme droughts, a rising ocean, or floods. The financial community's surging interest in assessing and understanding climate risks builds on this new understanding of the broad capital base.[35] There are little or no profits in a disintegrating society suffering severe wildfires and climate breakdown.

Both the personal incomes we receive each year and the national incomes rely on the services provided by the stocks. Buildings provide shelter and comfort. Rivers and soils provide freshwater and food. The social capital provides trust, norms, institutions, lasting relationships, and predictability, which all provide the daily flows of social interaction and subjective well-being. Hence, we can view yearly income flows (of money, ecosystem services, and daily well-being) as a form of interest from our long-term accruing assets. If there is more income any year than we consume, then the surplus can become savings. The savings can be reinvested in either productive, natural, or social capital. Then our broad wealth will increase as we move from year to year, and from one generation to the next. "Investing in natural capital" is the accountant's cold way of saying we need to care for the land, to involve ourselves in the flourishing of other than human beings, to give back to nature.

But what if a company, city, or country starts consuming, rather than adding savings to, its productive capital? This leads to depreciation—crumbling buildings or bridges, customer goodwill, ailing power grids, or other infrastructural woes. Or what if it starts consuming the natural capital for short-term profits—cutting down forests more quickly than they can regrow or drawing down rivers, soils, and freshwater aquifers more quickly than they can replenish? And what if its leaders destroy their trust in institutions—by discrimination, ethnic exclusion, corruption, violence, bigotry, or tax dodging? Then the broad wealth (productive, natural, and social assets) starts a long-term decline, even if annual GDP may still grow for some years. In that case, a country's economy may *appear* to be doing well in terms of annual GDP. But that's only at the cost of overconsuming its broad capital reserves, undermining its own long-term wealth. It's like overspending on your savings account or credit card. It only works for a short while, and usually leaves you worse off than you were before.

So, annual flows depend on stocks. And stocks grow when the surplus from the flows are used for reinvestment. In a nutshell, income is interest on wealth (from the bottom row in figure 5.3 and up). Savings increase the wealth (from the top row and down). Therefore, the steady, balanced accumulation of all three of these capitals over time is the key outcome of healthy growth. For a company, this annual value creation shows (or should show) in the profit-and-loss statement, and its wealth should be reflected in its broad balance sheet that shows improving solidity over time. This kind of broader accounting is increasingly called "integrated reporting."[36]

The Inclusive Wealth Report 2018 is a large research effort initiated by the acclaimed economist Partha Dasgupta along with UNEP. It uses a related approach to assess the development in the broad wealth of 140 nations.[37] After detailed valuation and documentation of the period from 1992 to 2014, their main finding is that—for most countries—growth in GDP is actually overconsuming natural capital in particular. But many are also depleting their human capital. Hence, they found that 44 out of the 140 countries have suffered a *decline* in this broad inclusive wealth per capita since 1992. This happened despite *growth* in GDP per person in all but a handful of them. These countries are not on a path to healthy growth even though their economies, as measured by GDP per person, are growing. In other words, despite all the frantic economic (gray) growth, they were less wealthy in 2014 than in 1992. They are depleting their stocks of natural, human, or productive capital at rates that will leave future generations worse off.

What is needed, then, is a change in the *direction of growth*. Not just ever more growth in incomes (GDP) and in the productive capital as priced in financial markets but a move toward a growth model that eliminates resource wastefulness sufficiently to bring down the environmental footprint, and thus kickstart real green growth. Over time, that can accumulate as savings to natural capital. Then, *at the same time*, we need growth that allows more and more of the population to share the benefits. This includes access to better education, health services, skilled jobs, and safety as well as gains in trust in and predictability from laws, government, the commons, and many communal institutions. The key to growing social capital is to keep inequality at the relative low levels conducive to trust and well-being.

Healthy growth, then, means changing the direction of growth from happening only in the leftmost column to include the two others. Not only growth in GDP and labor productivity. But also growth in natural capital (through sufficient resource productivity) and in social capital (through sufficient *social productivity* of growth, a concept I'll define in chapter 7). In figure 5.3, this is illustrated as a lateral direction of growth from the left column to the two to the right. When successful over time, this can grow our broad wealth, not just our annual economic income flows. With healthy growth over time, there can eventually be a healthy economy. Then we will have achieved a complete systems change by annual, verifiable steps of healthy growth.

It could even be time to drop the vague "sustainability" concept and replace it with this more specific healthy economy framework. "Healthy growth" is more exact than sustainability because it refers to measurable, sufficient rates of change in the resource and social productivity of the annual value creation (flows). And "healthy economy" refers to a socio-economic state in which the stocks of productive capital let people thrive within our ecosystem biocapacity in an equitable way that increases trust in society and well-being.

This change in the direction of growth can ensure that future generations have at least the same level of broad wealth as this generation. That is key, for sure. But even more important, strengthening the natural and social capitals, will also feed back over time into economic growth itself. There are many large, thorough studies that show how green growth investments can increase overall economic growth through new jobs and cost savings (by slashing the current wastefulness) and start regenerating well-functioning natural capitals. Countries, cities, and citizens all thrive better with cleaner air and fresh waters in the surrounding forests, and recovering oceans.[38]

And finally, it is increasingly clear that greater equality is becoming a stimulus of economic growth, too, not a deterrent to it.[39] That's why we'll now turn our attention to the third column—more closely examining the social capitals, with specific attention to the surprising costs of increasing social *inequality*.

6 INCLUSIVE GROWTH OR GROWTH ONLY FOR THE LUCKY FEW?

In 2018 I was visiting Cilincing, one of Jakarta's most polluted areas. Where the slow-moving river meets the ocean we—a bunch of Norwegian lawmakers—watched a group of children play. Amid all the rubble, the bamboo shacks, the stench from the shore, and the countless plastic bottles amassed behind them, these eight- to ten-year-olds bounded back and forth and radiated something altogether incongruent with their surroundings: joy.

We were on a mission to strengthen Norwegian–Indonesian collaboration on the oceans, plastic debris, energy, and rainforest preservation. Jakarta is one the biggest urban regions in the world, approaching thirty-five million people. By the late 1990s, the Cilincing area had become an industrialized port and a garbage-strewn slum. Up until a few years ago, we were told, you couldn't even *see* the water where the river met the harbor. It was obscured by floating plastic debris so dense you could, apparently, walk on it. Local fishing boats kept getting entangled in the mass of floating junk. Recent successful cleanup efforts and plastic recycling have made the river water visible again.

It was a weekday, but the kids weren't at school, if there was one. As they played, they showed off their vast cache of one-time-use plastic bottles, collected from the endless supply still sailing down the river to Cilincing. Soon, the bottles would be returned for cash, or for public transport tickets through the Plastic Bank project.[1] The kids were effective collectors.

One girl, Patimah, was watching my group with an amused twinkle in her eye. My Norwegian colleague wanted to pose for a selfie with some of the boys. So Patimah rushed in, organized the others, put herself in front to make the most incredible faces, framing her smile with her forefingers and thumbs. All her friends followed suit. We simply had to get our smartphones out. They were no strangers to what the tall, white, weird foreigners seemed to want (photo shoots!). I was aware of a certain ambiguity: Are these smart kids having fun with us—or making fun of us? Maybe both? My colleague was soon crowded out in the posing game, disappearing behind a bunch of laughter, waving arms, his voice drowning in the chatter.

Having fun posing, getting ice cream from the organizers of our trip, running around with other kids, chattering away—they seemed just fine. Healthy kids. Embodying the future, despite the surroundings. But, as has often happened when I travel through troubled areas, while smiling with them, I kept feeling the borders of despair. What will the future bring them as they compete for space and resources with the seventy million other young Indonesians currently below fourteen years of age? Sweatshop working conditions with long hours and low pay? Sex trafficking or domestic abuse? Informal settlements with no land rights? Health crises with no insurance? Will they fish for the last fishes before being completely driven away by marine debris, ocean acidification, and pollution? Will they be chased away by private security contractors, or militias?

As a white, senior male who was born in wealthy Norway, I am keenly aware—particularly in moments like these—about the level of privilege into which I've been born. I've never had to deal with the kind of destitution that dominates in these areas. But even though I've traveled far from Cilincing since, I still wonder where and how Patimah will be in ten years as she passes into her twenties. The same questions keep coming to mind: Will she manage to break out of poverty by excelling in education or entrepreneurship? Or will she already have three kids? Will she and her friends stay in abject poverty? Or will a tiny drop of some of the incredible economic growth in Southeast Asia at long last "trickle down" to her as well, so she can crawl out of the lowest global decile of income?

SOCIAL PROGRESS: THE SILVER LINING OF GRAY GROWTH

Let's start with the good news. The optimists are right: This is the best time ever to be a human. As both psychologist Steven Pinker and physician Hans Rosling recently illustrated, long-term trends point in the direction of better health, more literacy, greater longevity, and less absolute poverty.[2] Looking at the world through their lens, the prospects for Patimah and her friends are better now than they were in the 1980s, when billions of people were living in extreme hardship and economic destitution. Since then, living standards and many social conditions across the world have mostly, on average, improved.

It can be useful to look at those improvements against the backdrop of the Sustainable Development Goals in the United Nations' 2030 agenda. Consider just the first four SDGs: no poverty, no hunger, good health, and quality education. A research team I was part of analyzed seven major economic regions (either countries or groups of countries with similar GDPs and purchasing power): the United States, other rich countries, China, emerging economies, the Indian subcontinent, sub-Saharan Africa, and the rest of the world.[3] We found truly impressive advances for billions of people over the last thirty-five years everywhere we looked. In all seven regions and across each of the four goals, things moved in the right direction from 1980 to 2015. With increasing GDP per person, the world has laid the foundation for solving poverty (SDG 1) by lifting ever more of the population above the absolute poverty limit of $1.90 per day. In the same vein, hunger (SDG 2) afflicts far fewer today than forty years ago, even if there is still a long way to go in most regions. In terms of good health (SDG 3), almost all regions are now entering the midrange longevity. And in education (SDG 4), a majority of regions now give most people more than twelve years of expected schooling, which is in the green target zone.

The very strong patterns that emerge when plotted against GDP per person make a strong case for economic growth: Whether in Asia or Latin America or Western countries, economic growth per person has coincided with improving basic living conditions in the 1980–2015 period. Nearly all regions follow the same patterns: As countries and regions approach and eventually cross the threshold of $40 in GDP per person per day ($15,000/ year),[4] they move closer to eliminating poverty and hunger as well as delivering on education and health. If all regions manage to cross the $40 per person per day threshold by 2030, there is high likelihood that they will all enter into the "green" zone of goal achievement.

Continuing on the current course, by 2030 the world will hit the target of full elimination of hunger, absolute poverty, and better health and education—with the exception of sub-Saharan Africa, which would hit the target by 2050. All this, with just continuing with business as usual. As Hans Rosling pointed out in his book *Factfulness* and in his many TED talks, most people in the rich world have no clue about the extent of these recent good developments.[5] The mainstream media, he noted, overlooks this major but slow-moving good news.

Several more of the seventeen SDGs are on positive trajectories, too. Some measures of gender equality (SDG 5) are showing great improvement. The balance of boys and girls across expected school years is already almost one-to-one, which means that in most places girls have caught up on access to education. And in some regions (the United States, other rich countries, and emerging economies), girls now actually outnumber boys in school life expectancy.

Access to safe water (SDG 6), another key condition for dignified lives, has also improved in all regions. Lack of safe water implies a huge burden on women in particular in many poorer cultures and countries. The fraction of people with access to safe water was more than 80 percent on average by 2015, though sub-Saharan Africa still fell below 60 percent.

Giving people access to affordable and clean energy (SDG 7) is similar to that of giving them safe water: The richer world already delivers electricity to nearly 100 percent of its citizens. Both China and India are now increasingly delivering electricity to their billions of inhabitants. One could certainly dispute whether the majority of that energy qualifies as *clean* yet.[6] But solar power—both off-grid, microgrid, and on-grid—is rapidly changing this for the better. The number of people without access to any electricity is falling everywhere.

So, whether one "likes" GDP or not, growth in GDP per person still seems to be an effective way of funding improvements in poverty, hunger, health, education, gender equality, water, energy, and jobs. But mostly when starting from a low level. As GDP per person grows higher than $40 per person per day, the improvements tamper off. There are "diminishing marginal returns," as economists are fond of putting it. Economic growth helps at the start but can only take you so far on each of the goals. There is much slower progress once the $40 threshold is reached.[7]

Why do the benefits of higher income taper off? One reason is that the further benefits of economic growth get heaped upon a lucky few among the citizens. When that happens, the GDP *per person* gives a very incomplete picture of the social situation. That's because the *mean* income per person can be quite different from the *median* income. In statistics on household incomes, the mathematical average (mean) may be skewed by the few extremely rich. *Median* income, the actual midpoint in a range of incomes, may be a better way to suggest what a "typical" income is. And a striking fact is that median incomes have improved much, much more slowly than the average, if at all, in recent decades.

Thus, progress on the *social* development goals (SDGs 1–7) has been moving in the right direction since 1980 for most countries and for regions as a whole. And it can be expected to move toward the targets for 2030 and possibly attain them by 2050, assuming that economic growth continues. But there is one social SDG where indicators have been pointing almost exclusively and consistently in the wrong direction within countries, and that is reduced inequalities. The trends in inequality within countries (SDG 10) have been moving away from that goal.

So while Patimah, the girl I met in northern Jakarta, may end up crossing the international poverty line and earning more than $1.90 per day, she may still get locked into relative—if not absolute—poverty for a lifetime. The subcaste of the stagnant 40 percent lowest incomes in many populous areas live just barely above the poverty mark—especially in countries with low average GDP per person. Earning a little more than $2 per person per day is very, very far from any guarantee of having any dignity at work or at home. Exploitation, vulnerability to abuse, weak labor rights, corruption, and dehumanization in mass production systems can all exist well along an average of $2 per day, which is technically "out of poverty."

In summary: Long-term growth in GDP per person solves hunger, extreme poverty, basic health, years at school, water, and energy in poorer countries. But it *does not* pull the poorest fractions along very well. It does not secure working conditions. And it does not by itself solve inequality. Rather, the conventional growth model is—in many countries—making inequality even worse.

Increasing numbers of scholars are pointing to rapidly worsening inequality as the deeper cause of several other disturbing political, societal, and economic setbacks as we move deeper into this century.[8] It also undermines the benefits from progress made on other SDGs.

THE DARK SIDE OF RECENT GROWTH: RISING INEQUALITY

For many decades after World War II, the conventional growth model worked pretty well, at least in many OECD countries. The benefits of growth were shared among the strata of society. Both the rich and the poor gained from growth. It worked well not just for capital owners but also for workers, employees, the public, and people in general (if not for our ecology or climate, as we saw in the previous chapter). At that time there were a number of clear links between economic growth and social welfare. When investment went up, labor productivity went up. When labor productivity went up, wages went up. When employment and profits went up, wages also rose. And as wages went up, inequalities improved. When employment rose, even government tax revenues grew—which meant governments could afford more services for the less wealthy. A number of positive feedback loops improved society overall. This was particularly true for many Western countries in the 1945–1980 period. The result was lower inequality along with growing incomes for all in so many consecutive years that it established itself as a "natural and normal" state of affairs in the culture. People started to expect it.

But since the mid-1980s, *these links no longer work*. Now, when growth points up, inequality tends to rise, too, and even further. Researchers are

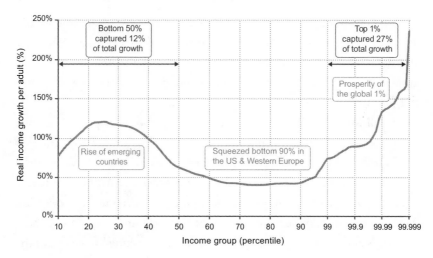

Figure 6.1

Income growth from poorest to richest adults for the whole world, from 1980 to 2016. Real income means market incomes (labor income + pensions + capital income) in constant prices before tax and transfers.

accumulating overwhelming evidence of this new reality from all over the world. The picture is quite stark: The richest 10 percent, and particularly the richest 1 percent, have rapidly captured excessive gains from growth for four decades. See figure 6.1 (from World Inequality Report).[9]

Inequality exists on two planes: income (a person's earnings from their labor or other sources, such as a pension) and wealth (the value of a person's accumulated assets). When we look at income inequality, we see that the world's top 1 percent of earners has captured more than *twice* as much income growth as the bottom half between 1980 and 2016. Luckily, however, the bottom 50 percent worldwide has nevertheless still seen some growth due to the rise of emerging economies. But the global middle class (which contains all of the poorest 90 percent in the European Union and the United States) has been squeezed.

Despite overall economic growth, the median income has stagnated in many advanced economies, such as the United States. Most Americans—the lower and middle classes—were not really any better off moneywise from the 1980s to 2015.[10] Not only has all the income growth accumulated in the top tiers, but the minimum wage has not kept up with inflation. Major living expenses, such as housing, healthcare, and higher education, have also gone up well beyond inflation. Slowly, this makes the lower income groups sink deeper into inequality and economic angst. They are "stretched thin," despite living in rich countries.[11]

In all human societies, people care greatly about inequality. For one thing, we are ultra-social animals. Evolution has deeply ingrained in our brains a tendency to worry about our rank in the flock. We scan the responses of our peers and worry about how others see us. But the harms of income inequality are more than just perceived, even in rich countries. As the value of individual incomes stagnates for a majority, so does that majority's ability to experience core benefits of a working life—such as the ability to feel secure, or to pursue education, entertainment, or a better standard of living. Hence the model of growth currently employed in these regions no longer works for most people.

The theory that money automatically trickles down from the rich to the poor has been proven wrong by recent data.[12] Markets are—by themselves—great at generating values but terrible at distributing outcomes evenly or fairly. Left unchecked, underregulated, and uncorrected, the accumulation mechanisms inside capitalism—which are exactly what makes it so dynamic and effective in production—will eventually turn it into rentier capitalism, which mostly serves the top 1 percent and particularly the one in ten thousand richest of the people. This rapid concentration of capital gains reminds us of the Gilded Age and the Roaring Twenties, two waves of inequality that characterized the decades leading up to the two world wars.[13] Similar accumulation has been happening over the last decades in many countries (but not the Nordics, and less in Europe than in the United States, China, or India).

Once again, how we measure a problem impacts how we discuss and analyze it—and ultimately how we solve it. For over a century, economists have typically relied on a metric called the Gini, named after Italian economist Corrado Gini, to measure the extent of income inequality in a city, country, region, or the world. On the Gini scale, inequality is measured between 0 and 1. If a region's Gini is 1, there is absolute inequality (one person gets it all, the others get nil). If Gini is 0, all citizens have exactly the same income. But it turns out to be difficult for people to understand a Gini point of, say, 0.328, the 2018 Gini score for net income in the United Kingdom.[14] Drawbacks include the obscurity of how it is calculated and its inability to describe what matters most: how the poorest and the richest compare. So the wisdom of using Gini as the standard measure is being rethought.[15]

One newer, better measure of inequality is therefore gaining ground—the Palma ratio, named after Chilean economist Gabriel Palma. He found that middle-class incomes very often represent about half of gross national income while the other half is split between the richest 10 percent and the poorest 40 percent. Surprisingly, it is the shares of those two outer groups that vary most across countries.[16]

Hence the Palma ratio effectively highlights the relationship between richest and poorest—the top and the bottom of the pyramid. The Palma ratio is calculated quite simply: the share of national income captured by the richest 10 percent is divided by the share captured by the poorest 40 percent. If the 10 percent take more of total income than the 40 percent, then the Palma is higher than 1. The share going to the middle 50 percent is usually more stable and hence less important for understanding the most dire consequences of inequality—the increasing marginalization, and expansion, of a nation's poorest.

Wealth Inequality: Even More Severe Than Income Inequality

Over time, income inequality builds up wealth inequality. Poorer people need to spend all they've got just to get by. Hence there is little to save. Those with very high incomes can save much more of their incomes, making it possible for them to create wealth through investments and accrue assets like real estate that in good times rapidly build value over time. These funds can be leveraged and reinvested. Capitalism then tends to do what it is programmed to do: it multiplies that capital (in "good" times). Wealth escalates for those among the rich who win in the competition for higher returns. Their rents and capital income soon supersede their earned labor income. In the early industrialized countries, such as the United Kingdom,

United States, and France, this happened in the early 1900s: There was rapid concentration of wealth to a few successful capital owners.[17]

Then, after two world wars and the Great Depression came a long economic boom, ushered in by Keynesian economic policies with progressive taxation. During that boom, the wealth share of the 1 percent richest declined slowly from over 40 in 1940 to around 20 percent of total wealth in the 1980s.[18] "Slowly" since in real terms, the growth of the pie was so big that the rich got much larger slices even while sharing more of it. It was a win-win. Since 1980, though, the wealthy are once again claiming inordinately—and from a social perspective, dangerously—large shares of the pie, especially in Russia, China, and the United States.

Also, globally, the "lucky few" (the one-in-a-hundred-million richest people, comprising a class of seventy people) control unimaginable sums of wealth in a financially globalized world. According to some estimates, by 2018 the twenty-six richest people in the world owned as much wealth as the poorest half of humanity, amounting to almost four billion people.[19]

Even in the otherwise rich United States, an incredible 78 percent of people today live paycheck to paycheck.[20] New words are developing to describe these workers: no longer the proletariat but the "precariat" because they live fragmented, economically precarious lives, always just on the edge of ruin.[21] Their plight has birthed the new acronym ALICE (Asset Limited, Income Constrained, Employed), coined by the charity United Way.

As the United Way points out, ALICE can be "your childcare worker, the cashier at your supermarket, the gas attendant, the salesperson at your big box store, your waitress, a home health aide, an office clerk. ALICE cannot always pay the bills, has little or nothing in savings, and is forced to make tough choices such as deciding between quality childcare or paying the rent. One unexpected car repair or medical bill can push these financially strapped families over the edge."[22] Despite working, often at more than one job, ALICE earns far too little for an economically secure lifestyle. No matter how hard these members of the precariat work, they're not going to "make it." Their kids and their communities will suffer in the long run, too.

There is no lack of money, capital, food, or material products in today's huge, global economy. It's just all extremely unevenly distributed and getting worse by the year. And now, the dawning age of robots and AI can replace a lot of low-paying jobs. How can capitalist societies deal with this escalating precariousness, economic anxiety, and social injustice—before it further undermines the trust and fabric of society itself?[23]

Any in-depth understanding and description of capitalism will point out why and how it increases inequalities in outcomes. In economics this reality is known as Okun's big tradeoff: either efficiency or equality.[24] In other words, capitalism begets inequalities from a mix of sheer capital accumulation mathematics and disproportionate winner-takes-all market gains for players with the highest productivity. Neither greed nor evil plots are necessary to ensure the outcome. Clearly, it's not the case that capitalists of the 1960s and '70s were kind and altruistic while those of the 1980s to 2015 had somehow turned greedy and selfish.

It's better to acknowledge from the start that such capital gains are an inevitable and predictable outcome of any type of unregulated capitalist system. If governments choose to weaken the equitable framework around it, then high inequality bounces back. Inequality is not a bug in capitalism, it's a feature—as software people would say. But if that feature is ignored and capital is left to accumulate by itself, the consequences are dire, as we'll see below.

In this sense, inequality is a public choice.[25] The choice is on the one hand to let capitalism predictably build high, costly, and unhealthy levels of inequality. Or on the other hand to employ the magic of well-designed labor and capital markets to create larger income flows that can then be redirected to include everyone.

THE COSTS AND CONSEQUENCES OF INEQUALITY

The more economic researchers dig into the increasing inequality in recent decades, the more they discover the severity of its consequences on human health, skills, life expectancy, and other social factors. Even mainstream economic organizations such as the International Monetary Fund (IMF), the OECD, and the World Bank have issued major publications and are undertaking research and other initiatives to alert leaders to the dangers of not addressing the problem. Among their main findings is evidence that increasing inequality weakens the level and durability of growth itself, and thus imposes a direct economic cost on *all* citizens.[26] In other words, inequality not only ensures human suffering in lower classes and poor communities, it also ultimately jeopardizes economic growth itself.

The links from inequality to slower growth seem to work through a number of pathways. Increasing inequality concentrates power with the wealthy and thus weakens the support for reforms and threatens political stability. High inequality also weakens human capital over time, as the poor are increasingly unable to fund good education and social mobility worsens.[27]

Long unemployment spells hit the already poor disproportionally and eventually lead to loss of skills. When people go without jobs for long periods of time—even if they have more than twelve years of education—they subsequently may only find sporadic, precarious, temporary employment. They can end up in a vicious cycle of being perceived as unemployable and staying in poverty. Their potentially valuable contribution to society is wasted.[28]

When inequality is mixed with loss of access to financial services, this weakens an individual's ability to cope with job upsets or financial, medical, or other risks. If widespread in society or exploited by credit sharks, this loss of access can exacerbate macroeconomic instability.[29]

High inequality may also weaken opportunities for marginalized people to join the economy, as it aligns with and reinforces both racial and gender discrimination and women's job access. Possibly as a repercussion, inequality may cause more social conflict, weaken trust in society, and thus weaken institutions and governance.[30]

As if all this wasn't bad enough, the economic research is joined and strengthened by findings from sociologists, psychologists, and health researchers who are extending the understanding of inequality's wider costs. They trace a number of other social problems, including mental illness, violent crime, imprisonment, teenage birth rates, obesity, drug abuse, higher rates of low birthweight, and worse educational outcomes for schoolchildren.

The tendency for more unequal societies to have higher murder rates has also been seen many times.[31] The rates of early death for poor men in otherwise equally wealthy countries rise with inequality. In wealthy countries, whether the United States or Norway, the poorest 1 percent of men live a whopping fourteen years less than the richest 1 percent.[32] Studies also document that specific forms of mental illness, including depression, schizophrenia, and psychotic symptoms, are more common in more unequal societies.[33]

In order to bring together this broad span of findings, researchers Wilkinson and Pickett constructed an index of health and social problems consisting of ten such problems and correlated that scale with inequality across different states and countries. The surprisingly strong correlation came out as shown in figure 6.2.[34] Such high significant correlations are rarely found in real-world, large, and messy data sets, so it really gets the attention of nerdy social scientists like me.

These researchers point out that by 2015 there were as many as 300 peer-reviewed studies on the relationship between income inequality and measures of health or homicide. Together, the studies provide overwhelming evidence that greater inequality is not just linked to but actually *causes* worse health, shorter lives, and more violence, something economic research has mostly omitted.[35] The causality seems to operate through its fundamental effects on the quality of social networks. When a person's relationships to coworkers, neighbors, friends, and others weakens—and when lower income brings lower status, isolation, and economic anxiety—a number of psychopathologies and social problems can take hold.

Inequality is thus in many ways similar to resource wastefulness: The gray, unfair growth model wastes not just resources but also human opportunity and skills by excluding them from gains in the labor market. We get thrown-away people, along with thrown-away plastics, nutrients, and carbon emissions.

Fixing the resource waste would still leave us with the inequality problems of growth. Even with the kind of rapid digital and resource

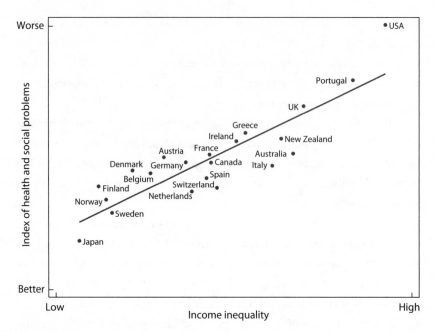

Figure 6.2

The surprisingly strong relationship between inequality and an index for health and social problems. Wilkinson and Pickett found the strength of correlation to be $r = 0.87$, and statistical significance $p < .001$.

productivity improvements described in chapters 2 through 4, we may still end up in unstable, violent, and bigoted social conditions. The technical developments work better when matched with long-term reforms in society. And it's not just the conventional vices of the human mind we need to worry about. The extra pressures of climate disruptions on existing social fault lines and weak governance systems will amplify the problems enormously. Why should the rich, who can be self-supplied by solar power, air conditioners, and fire-safe properties in gated communities on higher grounds, care for those vulnerable to abrupt and extreme weather disruptions? Such challenges will worsen as the inequalities grow and become ever more glaring through better social communications. Any have-not with only a dirt-cheap smartphone will eventually know everything about how the have-mosts swim in luxury.

In short, the capitalist engine, if left on its own, will necessarily increase income and wealth disparities. Unless there are interventions and institutional frameworks to rebalance them, we will arrive at a point where the injustice and destitution suffered by the have-nots become unbearable for all. And particularly so in periods of crisis. If so, how does civilization deal with this before the social fabric ruptures? In Europe a century ago, runaway inequality contributed to the rise of a lot of destructive -isms (like communism and fascism) and massive wars. What are the possible solutions? Does the answer lie in philanthropy or more progressive taxes? In shorter working weeks, with universal income or in "helicopter money" to the poor?

SOLVING INEQUALITY

There is a long, complex, and ugly history of kings and central governments trying to grab or claw back some money from the richest few. Leaving aside the thumb screws, guillotines, confiscations, pitch-fork uprisings, and governmental expropriations of the past, there is a more modern, large, and well-tested toolbox available for politicians to achieve a somewhat more egalitarian society—*if* there is sufficient political will to reduce inequality by choosing to use these tools. The standard choice of governments, and the most well-documented tool, is of course to legislate higher income and wealth taxes that increase as the taxable amount gets higher. These progressive tax rates are based on the principle that those who receive the most from society's markets should contribute the most back to society.

The main tools and solutions for dealing with inequality are listed in table 6.1. These are the policies and instruments that can create a more inclusive growth and a more egalitarian society. Some instruments work best at the top end of incomes and wealth, others at the bottom end. Some tend to design society to bring about less inequality in the labor market. *Proactive* tools are those that work pre-income—like trampoline policies such as job counseling and retraining to help people jump back in, or raising minimum wages. Other tools can reduce inequality by redistribution through transfers or the tax system. These are the *reactive* tools

Table 6.1

The inclusiveness toolbox

All the tools can be scaled up and down, and all have their pros and cons. The proactive/ bottom-wealth tools increase opportunities, thus potentially raising market incomes. The reactive tools are more about fair redistribution and reducing inequality in disposable incomes (net tax and transfers).

	Proactive tools (pre-income)	Reactive tools (post-income)
Changes for top wealth	Higher corporate tax rates Higher inheritance taxes Taxes on financial transactions Reform patent /Intellectual Property Rights (IPR) systems	Progressive taxes on income and wealth (higher rates/cutting deductions) Global tax registry and tax transparency Increase audit capacity of tax authorities
Changes for bottom wealth	Access to education, health Access to jobs (including job counseling and retraining) Collective bargaining/strengthen trade unions Worker representation on boards/employee co-ownership Secure small owner land rights/home ownership Real minimum wages/basic income/jobs guarantee Improve voting participation	Progressive taxes (low rates) Unemployment benefits Household transfers/cap childcare costs for low incomes Affordable (deductible) health insurance Higher Pensions Carbon Fee and dividend policies

that work post-income—correcting after the job and capital markets have already created high inequality. Pre-income means proactively designing for smaller inequality in market incomes. Post-income means correcting inequality in the tax system so that the net or disposable incomes after tax are less unequal.

Each instrument has its pros and cons and must be tailored over time to suit the culture and current trends. The grand political challenge is getting the mix right for each country while synchronizing taxes between states and countries. If done in a timely fashion, then the trust that constitutes social capital keeps growing rather than unraveling under the rifts of escalating inequality and crises. Policies for increasing investments in education, health, and access to jobs, for instance, are also direct investments in human capital. According to the genuine savings and inclusive wealth approach, human capital accounts for typically the most valuable asset of a modern economy, an astounding two-thirds of all national wealth.[36] The combination of increasing human capital (better education and health) and social capital (trust, stable institutions, diminishing inequalities) is also a powerful input to a healthy economic growth.

Therefore, rather than eroding the wealth of nations, or killing private-sector incentives for value creation, there is strong empirical evidence that these tools for inclusiveness can improve both the quality and the quantity of growth.[37] For instance, better staffing of tax authorities to increase the capacity for fair, efficient audits and tax transparency is one opportunity. Each dollar invested in better audit and oversight can return more than $100 in tax revenues, which can then be reinvested in human capital. This is particularly true for corporate taxation, where large loopholes require tax authorities to build considerable competence and capacity to audit in a fair, transparent way. Otherwise, those corporations that use a lot of resources to wriggle out of taxes through gray-zone initiatives get an unfair advantage in competitiveness relative to those who pay their fair "membership contribution" to maintain the social fabric. Tax dodgers may gain individually as free riders in the short term. But they too will lose out as overall economic performance lags over the long term, because the trust in society that composes social capital declines.[38]

The inclusiveness toolbox can reduce excess inequality and rein in the runaway unintended consequences of the capitalist turbo engine. But only given sufficient political will. In democracies there is—in theory at least—majority rule. Since the numbers of the poorest 60 percent far out-weigh the numbers of the richest 40 percent, there should be—rationally speaking—a majority in favor of taxing the richest more, with the proceeds going to the poorer. In a word: redistribution. But people commonly don't vote that way, at least not in the United States or the United Kingdom.[39]

It seems that Nordic countries have most consistently applied a broad range of these tools over the last century: Sweden, Denmark, Finland, and Norway. Recent studies find that their Palma ratios are around 1.0, which means that the richest 10 percent get the same total amount of income each year as the poorest 40 percent together.[40] That seems to be a psychologically healthy or tolerable level of inequality conducive of trust and well-being.

The ratio has not risen over time, due to the extensive use of tools to make the economy work inclusively. By comparison, the United States' Palma ratio was three times that of the Nordic countries (in 2014) in disposable, post-tax incomes. That means the richest 10 percent of Americans collect at least three times as much income as the poorest 40 percent all together.[41]

Maybe that explains why the Nordics are consistently ranked above the United States in terms of human development, happiness, and quality of life indices.[42] A common finding of cross-country comparisons is that the Nordics succeed better than other countries in combining economic efficiency and growth with a peaceful labor market and a fair distribution of income with high levels of trust and social cohesion.[43] Most of the tools above have been applied consistently over decades. Many observers around the world are amazed that, with such a large public sector "body" and relatively smaller private market "wings," this bumblebee can still fly. And that it flies so well! The Nordic economies seem to prosper and grow in spite of the presumably weak economic private incentives associated with high tax wedges, a generous social security system, free education, and a rather egalitarian distribution of disposable incomes, household transfers, and generous public services.[44] Why work hard if the state takes a significant slice of your extra income? Still, in the Nordics, people do that and achieve high efficiency. Together, these bumblebees are a flying contradiction of Okun's big tradeoff of efficiency versus equity.[45] They are both effective and equitable.

Across the world, citizens appear unaware of the extent of inequality, the full range of solutions available to deal with it, and how best to tailor those solutions. But the real question is why there is a lack of political will to use the toolbox at all. In 2017 in the United States, the Trump-led Republican Party gave the largest tax cuts to the richest. The top tax rate was lowered, the corporate tax rate was slashed, and large exemptions were given on the federal inheritance tax. This is the opposite of progressive taxes; they were highly *regressive*. It has worsened the nation's inequality levels, and will continue to do so if not reversed in coming years.

Just as the green growth compass can help us accurately describe and identify gray growth, recognize greenwashing, and measure progress toward growth that accommodates one-planet living, a *fair growth compass* can help us articulate a broad, more accurate perception of the economic growth dynamic, one that measures whether inequality is increasing or decreasing each year. Such a compass would help us determine the direction and speed of this ominous trend of the lucky few racing ahead, leaving the poorest majority in the dust, pitting poor groups against other poor groups, and gradually undermining the quality of growth and the stability of our societies.

THE FAIR GROWTH COMPASS

Let's get a bit technical again to see what a fair growth compass looks like. First, it consists of two main dimensions critical for achieving the SDGs in the coming decades: one to show the rate of change in GDP per year, the other to show the rate of change in inequality per year.

We measure growth as we did in our green growth compass, and we can measure inequality with the Palma ratio. This value can increase or decrease by some percent per year and will reflect how much the richest 10 percent are taking relative to the poorest 40 percent.

When combining the two dimensions, as in figure 6.3, we get a compass that shows what kind of growth we're dealing with for a certain country, state, region, or even city: Growing fairer, toward shared prosperity? Or heading toward more unfair growth, benefiting mostly the lucky top 10 percent? Or toward fair degrowth, with both inequality and GDP declining? Or an unfair degrowth, with falling GDP and the poor losing out more quickly than the rich?

The vertical arrow depicts GDP growth. Its upper extreme is marked by a 10 percent rise in GDP per year, and the bottom as a 10 percent decline in GDP per year. The horizontal arrow depicts the annual change in inequality, with the extreme right denoting a 10 percent per year rise in an economy's Palma scores, signifying rapidly worsening inequality. If Palma declines by 10 percent per year, as marked on the extreme left, then there have been large strides toward a more egalitarian society. Certainly, swings beyond plus or minus 10 percent are possible, but history shows

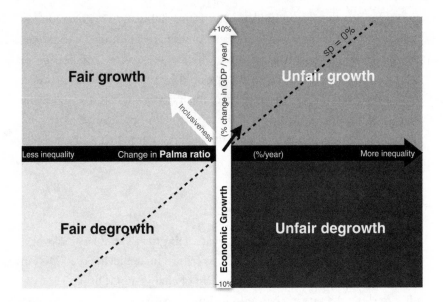

Figure 6.3

The fair growth compass: The vertical axis shows the change in size of the economic pie over a year. Larger is north (up), smaller is south (down). Moving east (right) on the horizontal axis means a larger slice to the richest 10 percent, moving west (left) means a larger slice to the bottom 40 percent. Changing course to the northeast means a more inclusive growth—a larger slice of a larger pie to the poor. The short, small arrow shows the global average annual change in 1993–2013. If the rate of change in GDP is equal to rate of change in Palma, there is zero inclusive growth (see the dotted line marked $sp = 0\%$).

that even dramatic years mostly fall within these parameters. For instance, when the United States was at the height of its tax-cutting, union-busting Reagan era, disposable income inequality increased by 9 percent in a single year, from 1983 to 1984. From 2006 to 2008, the Palma dropped by 3 percent per year, and hence the United States was moving into the fair growth quadrant—only to see the Palma ratio bounce back up 4 percent in 2008–2009 when the recession hit, pushing deep into the unfair degrowth quadrant.[46]

Unfair degrowth is also what happened in Russia in the 1990s as the Soviet Union collapsed: GDP fell, but some of the richest 1 percent got insanely rich. Inequality skyrocketed. Both the United States and Germany were in this quadrant from 2008 to 2009, in the Great Recession. Greece too was here since its 2010–2018 government debt crisis led to both falling GDP and higher inequality. We can also imagine contemporary Venezuela and Syria in this quadrant—without hard data backing it up, as, understandably, there are no real income data available.

For examples of fair growth over long stretches of time, we can look back to France, the United Kingdom, and the United States from 1950 to 1980: GDP went up, and the Palma ratio went down as inequality was reduced. The same countries then shifted to unfair growth from 1990 to 2015.

Finally, fair degrowth (the southwest quadrant) signifies a lower GDP but a fairer distribution of incomes. But there are few or no historic examples of it in the data. It seems that degrowth tends to come only in unfair versions, that is, always increasing inequality.

Together, these four directions represent the possible trajectories for a country's annual wandering. But they can also be used to make long-term scenarios on the future of inequality: In what direction are societies heading over the coming decades? And how can we measure their movement in a precise and objective way? The growth compass provides this measurement tool: Any country's course can now be charted with exactness (given that countries prioritize improving data quality and transparency on income inequality in the same manner as GDP accounts).

Degrowth advocates speak about the need for redistribution, rather than continuing the addiction to economic growth and GDP. To paraphrase: "Stop growing the pie. It's big enough, if we only share better what we've got." The arguments behind this reasoning are based on the type of gray, unfair growth that many OECD countries have seen over the last decades. There is no doubt that continuing with this conventional 1900s-type of growth will bring a huge set of worsening problems, both in terms of resource use and social inequality.

But the question now becomes: Does reversing growth into degrowth make these problems go away? Does it help societies deal with inequality if GDP drops, or dips into contraction for a while, or even long-term depression?

Even short-term recessions have longer-term consequences. First among these are higher unemployment, with some laid-off workers struggling to get back into stable, paid jobs. Affected families may be forced to put off saving or pursuing educational opportunities. Getting a loan to buy a new home, or even setting aside savings for a rainy day, becomes more difficult as incomes drop. Some rich people lose a lot of their wealth, too, but income inequality still goes up rather than down. The quality of life and standard of living for most people start to decline as well, which can affect health, overall well-being, and family stability. Businesses also start to feel the pinch. As consumers freeze their spending, small-business profits start to decline and large companies may not only cut jobs but also stop investing in the best available technologies—including those that could increase resource productivity—and research and development.

For any group or political party attempting to drive through a degrowth/contraction to improve environment and equality, the risk of a political backlash is severe. Most often, societies in contraction tend toward unfair degrowth rather than fair degrowth. For both rich and poor classes, it's most often a win-lose or lose-lose game.

We can apply this growth compass on a state or even city level as well. The consultancy Euromonitor has calculated the development of Palma inequality for many cities over the 2005–2016 period.[47] Among cities in developed countries, they found that the most unequal were Miami, Frankfurt, and Houston in 2016. In Asia, they found the highest income inequality in Kuala Lumpur. But the most unequal cities in the world were generally either in sub-Saharan Africa (such as Johannesburg, Lagos, and Nairobi) or the Americas (such Santa Domingo in the Dominican Republic). Indeed, of the twenty most unequal cities, sixteen were found in these two regions in 2016. But whatever starting point they have in absolute terms, any city or region can choose to set its goals and track annual relative progress in a better direction with the fair growth compass.[48] It's all about the *rates of change* going forward.

Could the fair growth compass be used at the company level, too? Yes, but not directly. It makes little sense to use the Palma ratio (which really only applies to all people living within one geographical area, just like the Gini). But a company's contribution to more inequality or a more egalitarian society can still be calculated. The relevant metric is labor's share of a company's total value added (gross profits). This share says something about the balance between workers and owners. The basic problem is that the labor share of income has been steadily declining since 1980,[49] and this decline is a root cause of increasing pre-tax inequality. In recent decades, under the slogan of "maximizing shareholder value," capital owners have taken more and more of the total cake at the expense of the workers' share.

The labor share of annual national incomes in the world's largest economies (G20) declined more than 5 percent over the 1991–2011 period (from 63 percent to 58 percent). In some countries—like Spain, the United States, and Australia—the labor share of national incomes has declined even more than 10 percent in the 1970–2014 period. For each passing year, this keeps adding to wealth inequality over time.

To demonstrate healthy growth, companies could choose to be open about how they contribute to improving or worsening inequality in society by disclosing the full distribution of their annual value creation. This is called a value added statement (VAS),[50] and it shows whether a company increasingly benefits the capital owners or its workers and the government (and thus society). Remember that value creation (= value added) is the total sales of a firm less all purchases of inputs from other firms. In US accounting, value added is the same as gross profits. What is left in the company's coffers is available for employee wages, investment, taxes, finance expenses, and finally owner dividends. When we add up the sum of *all value added in all enterprises* in the economy, we get national income.[51]

In its annual report, the Norwegian sensor-based sorting solutions company Tomra Ltd, for instance, shows how its value creation is split between employees (75 percent), the public via taxes (8 percent), shareholders (11 percent), and the financial sector via loans, shareholders in subsidiaries, and the like (6 percent).[52] Tomra thus contributes 75 percent of its value creation to the labor share of the nation's total income. Plus, most of the taxes it pays will go to funding public salaries. Since this level (75 + 8 = 83 percent) is way above the average share in the Nordics (55 percent), the more this company grows while keeping this distribution, this will push in the direction of less overall income inequality.

Thus, both the level and the trend are significant here. If the labor share of value creation (employees plus taxes) grows from one year to the next, a company with many employees is contributing more to fairer growth than the previous year. On the other hand, if the share to owners (shareholders plus financial expenses) is higher than the national average, then the company is contributing to inequality. And if the share to owners is growing year by year, it is increasingly doing so. This happens because larger gains to investors and lenders will automatically lower the labor share of national incomes. (If some or all employees hold some stocks in their company or through their pension savings, then a tiny bit of dividends may increase the income and wealth of workers, too.) But because wealth is already much more unequal than incomes, the better part of these returns on capital will nearly always go to the top deciles of wealth owners. Thus, all companies can contribute to healthy growth by first publishing their value-added statements as part of their annual reports, showing the distribution between stakeholders. This simple disclosure, combined with targets for inclusivity, could substantially accelerate the system change.[53]

FROM ANNUAL FLOWS TO INCLUSIVE WEALTH IN SOCIAL CAPITAL STOCKS

Solving the inequality crisis is as critical to healthy growth as solving the climate crisis. The good news is that when inequality is reduced, a whole series of other social goals benefit. It becomes a leverage point for further reducing poverty, malnutrition, and crime rates; improving education, longevity, and health; creating jobs; and more. It also lessens the barriers to growth brought on by inequality—which is why even the OECD, the international "rich countries club," emphasizes that we're "In It Together" as their 2015 report with that title states. Reversing the recent surges in inequality serves even rich people's long-term self-interest.

The magic bullet, however, does not reside in any single year's rate of change in GDP or Palma ratios. A high rate of inclusive growth for a single year is great. But focusing on short-term gains is quite myopic. The real challenge for a healthy economy is to ensure such rates consistently over periods longer than three to five years (in other words, decades). Each year with sufficient rate of change can be seen as a steady transformation of economic growth into better long-term social capital. That is real system change, year after year.

Building trust, institutions, transparency, justice, and cohesiveness, and other key factors in social capital, *takes time*. The public funds used are not lavish government spending but rather actual investment in our common, long-term social wealth. (It's just not visible on yesterday's accounts, as it's an externality without price.) Addressing inequality hinges on building up a broader base of capital for current and coming generations.[54] It's about becoming *better* capitalists, not killing capitalism. That's why political focus on growth benefits when shifting from maximizing short-term gains in annual economic flows to balanced long-term gains in productive, social, and natural capital stocks.

The purpose of both the green growth and the fair growth compasses is to help citizens, cities, and politicians visualize and connect the most important short-term steps, year by year, with a long-term destination such as 2030 or 2050. It can help us see whether we're getting the type of growth we want every year or whether we are fooling ourselves, from company to country, chattering about sustainability but wandering without knowing our actual direction. The critical questions to ask along the way: Are the changes in the *right direction*? Are the annual steps in the compass (the rate of change) *sufficient* to achieve a thriving economy with one-planet-compatible well-being and social stability in time? If anyone states publicly that "GDP is up 2 percent," to have a fuller picture we should ask: But how much are resource and social productivity up? Only then can we see if that growth is healthy or not.

7 HOW TO KNOW HEALTHY GROWTH WHEN YOU SEE IT

The Oslo Energy Forum is an invitation-only event for one hundred or so petroleum and energy executives, politicians, and select experts. There, CEOs from ConocoPhillips, Equinor, Shell, and others gather at Oslo's most expensive hotel alongside the Storting, the Norwegian parliament.

At the 2019 forum, the hot topic was the coming energy transition. For three days, the executives gathered there discussed the scope and pace of "green" energy technologies and the opportunities they presented for their companies. I was invited along with Johan Rockström, who was delivering a closing keynote on planetary boundary science. Leading into that keynote was a particularly interesting panel discussion on how fossil-fuel corporations could regain the public's trust.

That is a steep hill to climb, especially for those oil giants that spearheaded climate denialism through disinformation campaigns, continued business-as usual-operations long after they knew the dangers their products could cause and used dark money to keep governments in their corner. An increasingly distrustful public doubts their claim that they are the best bet for energy-transition leadership. Yet they also have the financial and technical resources to accelerate the transition—if and when they choose to do so.

The technology director of Royal Dutch Shell explained how much Shell was contributing to the transition by building some windmill projects and bringing "cleaner" energy (read fossil gas) to 100 million people in Africa. The CEO of fertilizer giant Yara explained that its fossil fertilizers are necessary to grow food for the world's billions of poor. No one on the panel disputed the need for transition. Rather, the presenters characterized their companies as publicly misunderstood providers of global energy security—stalwarts that are ready to step in and lead the transition to renewables when society is ready to quit fossils.

If ever there were a year when society could be deemed ready, though, 2019 was it. A wave of climate activism swept the world, several governments declared a climate crisis, scientists saw Arctic ice melt faster than anyone previously thought possible, and the brutal impacts of climate-intensified fires, floods, storms, heat waves, and droughts became ever more real to citizens. Yet the panelists stressed that their companies' efforts for the energy transition wouldn't fully kick in for ten years. Until then, not much would change due to the long lead times of large fossil infrastructure and long-lasting demand from existing fossil cars, factories, ships, heating systems, and the like.

As major players in the energy industry, those in the room held the power to either promote or stall the move away from their fossil products and toward renewable ones. I rose up to ask them: Why wait until 2030? Given the commitment to green business practices they had just described, would they be willing to publicly commit to a science-based target for improving their carbon productivity by at least 5 percent per year starting now rather than pushing it forward a decade? And would they also provide transparent public reporting to document that they were indeed doing their fair share today for reaching the goals of the Paris climate agreement?

The panelists responded by stressing how science-based their companies were and how much they supported the Paris Agreement. The message: They are the good guys. But they also carefully avoided committing to any sufficient minimum yearly carbon productivity improvement.

In the current climate crisis, though, winning slowly is the same as losing, and climate delayers are no better than climate deniers. It is very easy for a corporation to *talk* about clean energy and the need to tackle climate change, hasten the energy transition, ease inequality, or make green investments. And they do take a few positive steps here and there, like funding some wind projects or building a public school near a new gas field in Africa. Otherwise, it's business as usual while their communications departments get the job of hyping the windmills and the school all over their main markets, in social media, and on the web. This is the virus of greenwashing. Its symptoms include tongue twisting, confirmation bias, self-justification, and a continual cherry picking of facts. It causes some people to get sudden but illusory green flashes, and others a deepened cynicism and despair.

To cure both companies and citizens, a new treatment is needed— one that allows sufferers to not only know greenwashing when they see it but also to recognize genuinely green and healthy growth. This requires that we continuously clarify, objectively measure, visualize, transparently communicate, and broadly discuss this new growth model. But that also raises a host of questions. Should we judge performance based on the companies' achievement of their own chosen targets, regardless of whether they have set ambitious targets? Or should we judge them by their performance relative to other companies, so that only the best in class are truly green? Or is there another standard altogether?

THE STREETLIGHT EFFECT

The search for these answers reminds me of the old joke in which a drunk man paces under a streetlight, intently searching the ground. A police officer comes by and asks what he has lost. His keys, he says, and they continue looking together. Soon the officer asks if the man is sure he lost his keys in that spot. Whereupon the man says, "No," and explains he likely lost them in the park "but this is where the light is!"

Our tendency to seek answers where they seem easiest to find is age-old. So, what streetlights do we flock to when analyzing green growth?

The first is random targets. It is all too easy to get caught up in the "goals game." We find ourselves cheering any company that announces, say, a 25 percent cut in emissions by 2025. Or we applaud a country that announces a 30 percent cut by 2030. But do we know if those targets are high enough, or even how those exact goals came to be chosen? Did the matching numbers simply make for a good slogan? Seeking security in goal setting alone makes us fair game for the corporation that sets low-threshold targets that they're going to achieve anyway, or for the organization that adopts a few cherry-picked UN's Sustainable Development Goals, or even for the politician that backs hopeful but faraway targets without taking the action today that will ensure they are eventually met.

The second streetlight? Inscrutable rankings that measure corporate behavior in an effort to tell us which companies are the most sustainable relative to their peers. Again, the motive is good, but the reality of these rankings (ESG, i.e., environmental, social, governance) makes them yet another dangerous beacon for finding our way to a livable future. The sheer abundance of rankings, and of complex metrics being applied to make them, is so great that few really understand the math and weighting behind the final assessments. In reality, a company could top one list and rank low on another, and even those often ranked higher than others can still be undermining Earth's life-supporting systems.

Executives and politicians understandably want to select and present metrics that make them look good relative to others in the same industry, or relative to their own past. So Google highlighted its declining rate of CO_2 emissions per search—because it is easy to reduce that figure when the number of searches keeps escalating while servers get better. Equinor, the Norwegian oil producer previously called Statoil, emphasized that the CO_2 emissions from production in its newer fields are lower than emissions of oil companies elsewhere. Yet that has more to do with its location than its commitment to improve operations, because Equinor produces oil under Norway's much stricter regulations (including a ban on gas flaring). Airlines talk about emissions per flown seat-miles (which easily go down because of fewer empty seats and somewhat improved newer planes), and ignore the fact that their total emissions are still rising, due to rebound effects.

In fact, large companies have built sizable communication teams to make both their targets and their sustainability reporting look good. They can easily confuse any ordinary citizen with little time to spend critically examining the fine print. But this approach is, of course, a fertile breeding ground for the further loss of trust in corporations, capitalism, governments, and politicians. None of the panelists mentioned this in their discussion about regaining the public trust.

A FINAL COMPASS

Moving away from these narrow streetlights requires a better dashboard driven by a few key metrics that can be uniformly applied at many levels and are easily understood by laypeople, executives, and politicians. Toward that end, this chapter provides one more compass—the healthy growth compass, which builds upon the green growth compass of chapter 5 and the fair growth compass of chapter 6. At a glance, it shows whether a corporation, city, or country is truly moving in the right direction. Anyone can use it to evaluate their own progress, as well as to distinguish imposters from genuine healthy growth practitioners.

Like other sustainability measures in use around the world, the healthy growth compass relies on integrated reporting techniques that record impacts on natural and social capitals, not just financial ones.[1] Unlike them, though, the healthy growth compass addresses two critical questions: Is Company A or Country B's performance *sufficient* to start solving our shared challenges of one-planet living? And is it doing at least *its fair share, today,* of making the economy compatible with both a safer Earth and a more just society?[2]

In its simplest form, as seen in figure 7.1, the healthy growth compass is a three-dimensional tool showing eight possible growth trajectories. If you lead a growing company or city, there is just one course to follow for equitable, one-planet living:

Fair × Green × Growth → Healthy

Using the compass, you could chart the annual results to see if you are heading in the right direction now and taking sufficient steps each year to meet the (long-term) goals.

But, ahh, the goal. How does your company establish both green growth and fair growth goals in a thoroughly science-based way—one that doesn't allow you to hype your best metrics and downplay your worst, and

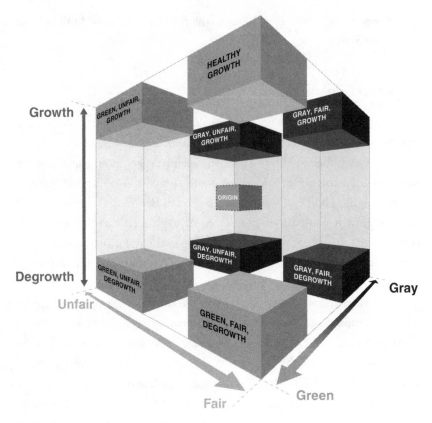

Figure 7.1

The three-dimensional healthy growth compass can be used to visualize the speed and the direction of growth for a company, city, region, sector, or country. There are eight possible directions, though the desired direction is toward healthy (i.e., greener, fairer) growth.

one that gives a true picture of whether your activity is helping or hurting people and the Earth?

Let's start with green growth and the planetary boundary science that Rockström presented to energy executives at the Oslo Energy Forum. Rockström led a team of scientists that in 2009 introduced the nine planetary boundaries that we discussed in chapter 5. The team illustrated those nine boundaries in a now-classic diagram (see figure 7.2). They also identified humanity's safe operating space within them, and plotted human impact in

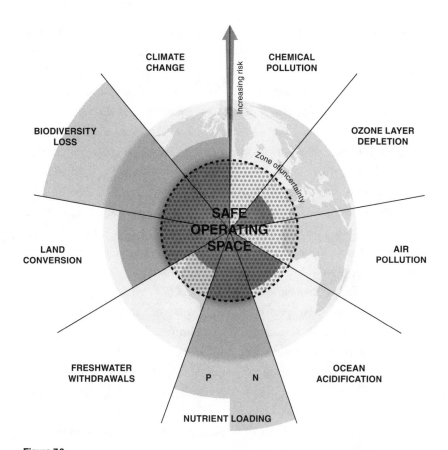

Figure 7.2

The nine planetary boundaries: The inner, dotted area represents the safe operating space. The dark shading marks humanity's impact on each. For instance, human activity has dramatically overshot the safe boundary for nutrient flows—even more for nitrogen (N) than for phosphorous (P). The greater the human-caused push outside the dotted circle, the greater the risk of large-scale, abrupt, and irreversible Earth system changes.

relation to them. Their warning: When human pressures push beyond the safe operating space, vast webs of life in our oceans, rivers, lakes, glaciers, forests, atmosphere, soils, and other ecosystems can begin to unravel. The consequences are felt across all national borders. And the more humanity presses some of these boundaries, like climate, into the high-risk red zone of uncertainty, the closer it comes to triggering Earth-system changes that are irreversible on a human time scale.[3] Due to uncertainty, we would only know with 100 percent certainty sometime *after* triggering it. That's why this approach speaks about risk levels: green (safe), yellow (boundary), and red (high-risk).

The planetary boundary model has come to define the Earth system framework within which the global economy can thrive at safe or medium levels of risks. The dark "pizza slices" show how far humanity has already pushed beyond the safe space on four planetary boundaries: climate change, nutrient loading, biodiversity loss, and land conversion, which includes forest loss. If human pressures are reined back into the safe operating space as soon as possible, if humanity starts regenerating the natural capital, perhaps the Earth will turn out to be resilient enough to recover. Then nature can possibly regain much of the benevolent and relatively stable conditions that characterized the last 7,000 years during which our fledgling civilizations evolved.

The cause of these pressures are the material flows of the human economy that unleash vast amounts of pollution from extraction to end use. The mega-corporations, cities, and countries with the largest, richest economies must do the most to annually improve resource impacts in absolute numbers until humanity is again inside its safe operating space. But all *must* take action at a fair, sufficient rate of change every year, at a scale proportionate to their share of the economy (their value added as a share of the GDP).

A shift toward genuine green growth requires so-called science-based targets[4] and measurements for all of the relevant human impacts, especially those four outside of the safe zone. To date, though, too few corporations and countries have set science-based targets founded on sufficient rates of change and consistently reported on those rates of change. That reality underscores the need for a new dashboard, which the green growth, fair growth, and healthy growth compasses provide.

To get a sense of how we can determine a sufficient rate of change for a planetary boundary, we will use carbon productivity as an example and the green growth compass as our tool. While insufficient alone, carbon productivity often gets the highest priority in any green shift—not just because humanity has so precariously overshot its climate boundary but also because carbon emissions typically make up between 60 and 70 percent of the total ecological footprint.[5] Carbon footprint correlates well with the material footprint, too.[6] Carbon productivity makes a good starting point for other reasons as well: Cutting emissions helps curb long-term biodiversity loss, deforestation, ocean acidification, food insecurity, migration pressures, and unequal access to clean water and air.[7]

The rest of this chapter becomes somewhat more technical. The impatient changemakers who want to jump into action may want to skip to part III, and then return to this chapter later when they want to know how to measure and report progress in a more precise, objective, and comparable way.

Measuring Value Creation and Emissions over Time

In order to navigate toward healthy growth, we need measurements of both value added (gross profits) and greenhouse gas emissions from the same activities and scope. Economic *activities* refer to those that are entered into the bookkeeping accounts measured in money, while *scope* refers to the emissions generated by these activities, first of all the direct but also the indirect emissions from the whole value chain of the products, measured in tons.[8]

As chapter 5 explained, carbon productivity is the value added divided by the associated greenhouse gas emissions for any period. It measures how much value we create for each kilogram of greenhouse gases emitted. The global average carbon productivity was 2.2 $/kg in 2015. In the US economy it was 2.7 $/kg (GDP/kg CO_{2e}), and in the EU 4.1 $/kg. China had 1.4 $/kg while Sweden had 6.6 $/kg.[9] The global average carbon productivity needs to grow by 2050 to at least 12 $/kg for the 2°C target, or preferably 24 $/kg or higher for the 1.5°C target. To get some perspective on those numbers, if governments set a price on greenhouse gas emissions of around 0.2 $/kg, that would take us a long way toward those targets.

At the national level, the total value added sums up to the GDP. While GDP has many critics and its statistical limitations and other flaws are well known, it can still be useful for green growth policies as an indicator of overall economic activity levels.[10] The coming few decades are critical, say Earth system scientists, for shifting the economy toward staying within planetary boundaries. We don't have the time to reinvent a totally new global national accounting system or a completely new capitalism. So in these critical decades, it is highly likely that real GDP[11] will continue to be one of the dominant national metrics in practical use. Why not, then, take a pragmatic approach to see how this common metric of value creation can be better applied in monitoring policy, strategy, and societal change?

In recent decades, global GDP has on average grown approximately 3 percent per year. Europe, the Americas, and Africa have tended to be below 3 percent, while Asia has been higher. Moreover, global GDP growth (adjusted for inflation) is down from approximately 5 percent during the 1960s, when productivity increases in developed nations were higher.[12] This is largely due to greater labor productivity potential in improving primary (agriculture) and secondary (manufacturing) sectors at the time, relative to the lower productivity improvements in the tertiary (services) sectors, which now make up about two-thirds of Western economies' GDP.

But how will global GDP grow toward, say, 2050? Moving forward, the year-on-year change in real GDP per person tends to decline as countries get richer. In the coming decades, we can therefore be pretty sure that average global growth rates will keep declining from the 3–4 percent range to the 1–3 percent range.[13] That's surprising good news from a climate perspective. It sets a cap on the growth of the total world economy, particularly as population growth slows, too, partly as a consequence of higher GDP per person. Going from 2050 toward 2100, GDP growth may fall asymptotically to near zero, as it becomes increasingly irrelevant to human well-being. That is the explanation of the flattening top of the S-curve in figure 5.2.

Measurements of aggregated greenhouse gas emissions are, like measurements of value creation and GDP, statistically demanding, both at company, city, and country levels. Science-based targets for carbon productivity are built on the global emissions pathways deemed necessary by current peer-reviewed studies for achieving—at the very least—the below 2°C target for global warming set by the Paris Agreement. Those emissions pathways are in turn based on the remaining global carbon budget as described by IPCC's latest consensus reports. Recent estimates of the remaining carbon budget for a greater than 66 percent probability of staying below 2°C range from approximately 600–1200 billion tons of CO_2.[14] Picking a midpoint of 800, and using 2020 annual emissions of nearly 40 billion tons of CO_2 as a guide, that leaves around twenty years on the budget. This indicates the need for pathways that reduce global emissions by more than 2.3 percent per year, at least halving emissions between 2020 and 2050. Ideally, to minimize risk and reach the vastly more preferable 1.5°C target, goals should be aligned with the lower estimate of the remaining carbon budget. To do that, we need a 6–7 percent reduction of emissions each year from 2020, resulting in around just 5 billion tons of CO_2 emissions per year by 2050.[15] Very ambitious, but not impossible.

Importantly, the entire range of allowable carbon budgets depends on the necessity of bending the global curve of emissions downward as soon as possible. Luckily, as Project Drawdown (mentioned in chapter 2) highlights, there are more than 100 proven solutions that can profitably start to reverse global warming before then.[16] We just need to grow and scale them up everywhere as soon as possible. The rates mentioned here are only the bare minimum. Anyone who's alarmed and inspired can lead by aiming much higher for their companies, communities, cities, and countries, surfing the sixth innovation wave toward better profits and lower resource use in the coming decades.

Climate scientists and economists have repeatedly calculated estimates of the minimum rate of carbon productivity growth to achieve the 2°C target. The *New Climate Economy Report* concluded that "the carbon productivity of the world economy (defined in terms of US$ of world output/ tons of GHG emissions) would need to increase by about 3–4 percent per year until 2030. In 2030–2050 the improvement in carbon productivity would need to accelerate again, to around 6–7 percent per year, to stay on track." The Deep Decarbonization Pathways project chose a model where the carbon productivity grows in steps per decade from 2 percent per year in 2010–2020, to 3.4 percent in 2020–2030, 5 percent in 2030–2040, and a whopping 8.5 percent in 2040–2050.[17]

How do we apply these targets to real green growth definitions and measures for companies, states, and countries? Green growth implies a resource productivity rate greater than the growth rate, or $rp > g$ as described in chapter 5. But some types of green growth are too slow to meet the global target of below 2°C (and certainly for 1.5°C). To tie resource productivity to rate of change to the planetary boundary of climate change, I'll introduce a new term, CAPRO for carbon productivity:

CAPRO = value added / greenhouse gas emissions per year, in $/tCO_{2e}$

capro = the rate of change in CAPRO in percent per year

A global average capro higher than 5 percent yields a reasonable likelihood for reaching the Paris Agreement's 2°C target. Despite variations in approach for these particular studies, many converge near that rate of change if we start now and rely on the higher end of the global carbon budget.[18] The same result arrives directly from doubling global GDP while at least halving emissions—yielding an average 2.5 percent GDP growth while reducing global emissions by at least 2.5 percent per year from 2020 to 2050. It is important to recognize that this 5 percent rate is very likely a bare minimum. In order to deliver on the 1.5°C target of global warming, capro must rise considerably (more than 7 percent per year). We are unlikely to keep below either 2°C or 1.5°C unless everyone starts to deliver soon on average above at least 5 percent. Not just any $rp > g$ is sufficient for those global targets.

The simple definition of genuine green growth (GGG) then becomes:

GGG requires capro > 5%

Attaining genuine green growth requires advances relative to other planetary boundaries, too, of course, but this is ground zero. Carbon productivity is a core component of the broader concept of resource productivity (RP) and its rate of change (rp).[19] Hence, a more general requirement of genuine green growth for all planetary boundaries in the high-risk zone can be calculated as:

GGG is: $rp > 5\%$

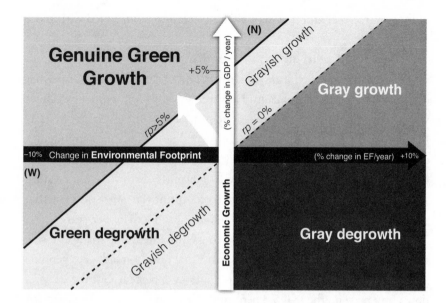

Figure 7.3

The genuine green growth compass. When the rate of change in resource productivity, Δ(GDP/EF), is more than 5 percent per year, we move northwest and enter into genuine green growth territory, a trajectory indicated by the thick arrow labeled "*rp* > 5%." Environmental footprint (EF) can be measured in tons of greenhouse gas emissions, domestic material consumption, and/or global hectares of biocapacity per year.

We can now update the green growth compass from chapter 5 to include science-based targets for green growth rates. To be healthy, all economic activity must shift direction toward the northwest quadrant by at least 5 percent per year on average as figure 7.3 illustrates.

When economic growth comes with a larger environmental footprint (whether measured as carbon, material, or ecological footprint), we get gray or grayish growth, which lies northeast in the compass. This is the type of growth most countries experienced up to 2015. If the rate of change in GDP for any year is exactly equal to the rate of change in environmental footprint, then there is a zero rate of change in resource productivity. Or:

when Δ(GDP/EF) = 0, then *rp* = 0%

Changes along this dotted line do not even make a relative decoupling. Below this line, we find gray growth (a clear coupling of value added with more resource use) and gray degrowth (decreasing value added with more resource use).

When GDP starts to grow faster than the ecological footprint, rp improves, and we enter the grayish growth area of the compass—somewhat better but still far from sufficient. However, when EF substantially declines while GDP grows, rp will truly skyrocket. If the improvement in rp is higher than 5 percent per year for an extended period, we are stepping into genuine green growth territory. Only there can we find sufficient decoupling for genuine green growth.[20] A 10 percent growth in value creation combined with a 10 percent decrease in footprint (far northwest corner), means a 22 percent improvement in resource productivity per year.

Adding to the nuance of this compass are two interesting triangles inside the northeast and the southwest quadrants that are also marked as genuine green growth. If, for instance, GDP declines by 1 percent, but the EF goes down by 6 percent, the rp is 5.3 percent.[21] We are then in the triangle labeled (W), west of the diagonal solid line marked "$rp > 5\%$." As rp is higher than 5 percent, this type of growth meets the criterion for genuine green growth, despite a small decline in GDP. This direction could also be a constructive way for some countries to get back within planetary boundaries. It represents, possibly, a pathway that degrowth proponents can also support.[22] Countries following this trajectory would most likely be rich, advanced countries that go through a contraction in resource-intensive sectors of production and consumption while simultaneously investing heavily in resource productivity. Rich Norway, for instance, could rapidly scale down its petroleum sector while investing in energy efficiency and electrification.

The other green triangle labeled (N) up in the upper right, otherwise gray growth quadrant is also very interesting. A company or a country that has strong GDP growth (> 5 percent annually) combined with a small EF growth might find itself heading there. India, for instance, might achieve 7 percent annual growth in GDP for a while. If this growth is resource productive (for instance, by running mainly on solar power), then its EF might go up only 1 percent per year. Then the rp would be 6 percent.[23] Being higher than 5 percent, it qualifies as genuine green growth. So, if this were to happen, India would contribute to the global average improvement in resource productivity despite its footprint increasing a little in absolute terms.

This upper triangle (N) makes it possible for poorer countries to catch up within a genuine green growth framework. A global convergence pathway toward 2050 becomes possible: In it, poorer countries are encouraged to grow more quickly than richer countries to battle poverty, hunger, and inequality. With 7 percent growth, India's more than 1.3 billion inhabitants could rapidly improve quality of life and the social SDGs. This would double its GDP in ten years and expand its EF by 1–2 percent per year for a while. And this must be seen as fair, because the per capita footprint in India is currently very small.[24] Employing the cheap new solar and digital technologies, India could undergo its modest expansion in footprint while other richer countries (which have much lower GDP growth rates) reduce their footprint more rapidly. As India's GDP per person eventually crosses $20,000—probably in the 2040s—its GDP growth rate per person will start to slow down (< 3 percent) as its population stabilizes.[25] Then, India's footprint will start to decline—as long as it keeps growing its rp more than 5 percent. In this way, India can both grow out of poverty for its many millions and do its fair share of decarbonizing and dematerializing the global economy average.

Hence, we could envision this north triangle as a suitable growth model for all poorer countries, which both need and want to grow their GDP to bring their citizens out of poverty and hunger and give them access to lighting and energy. But they can increasingly apply new renewable, local, digital, and circular solutions to achieve their goals. While such poor countries (in N) may need to expand their footprint modestly for rapid genuine green growth, this could also be compensated by all the richer countries as they shrink their EF much more quickly than they grow their GDP (far west in the compass, illustrated with W).

Similarly, a company with rapid growth in gross profits (for instance, 10 percent per year), can be a boon for green growth overall as long as its $rp > 5$ percent. It can then outcompete and replace average grayer competitors, despite its footprint possibly increasing a little (say 2 percent, as that gives an rp of 10–2 = 8 percent).

It's Not Where You Came From, It's Where You're Heading

Viewed this way, growth is reframed from economic growth to resource productivity growth. It is no longer all about capital, labor, and technology but about the footprint from resource use as well. The criterion of >5 percent rate of change moves us past arguments like "There is no room for the poor countries to grow" or "The rich countries must cut their emissions and consumption first." Such disputes stalled global climate negotiations for more than twenty years. Countries were unable to agree about how to share the burden of limiting emissions or financing common solutions. Now we can use this simple, clear criterion to plan and to check whether any country, city, sector, or company is doing its fair share, year after year.

If a country's, city's, or corporation's *rp* is lower than 5 percent over time, it is clearly part of humanity's overall problem. If it grows its *rp* higher than 5 percent for longer periods, however, it is part of the solution by contributing to system change. It doesn't really matter whether it has historically been in the degrowth, green growth, or gray growth quadrant: It only has to make sure that *from now on* it shifts into the large triangle that is defined by the >5 percent rate of change. As the singer Ella Fitzgerald once said. "It isn't where you came from; it's where you're going that counts."[26]

Away goes the vagueness, the hype around green growth, and the targets set on a whim. Any organization, city, or government just has to do a couple of standard things that most do anyway: document and report their value creation (in municipal or national accounts and in profit-and-loss statements), and transparently document and report their carbon emissions according to standards given by the United Nations Framework Convention for Climate Change for countries and the CDP (Carbon Disclosure Project) for companies and cities.

If anyone claims that their products, jobs, companies, or countries are green, then check their growth in carbon productivity first. If it's above 5 percent over time, they are doing their minimum fair share of the systems change for a 2°C target. If it is slower than 5 percent, they're greenwashing. If it's well above 5 percent, they're for real! And if they want to become true heroes, they could commit to the 1.5°C target, and then deliver higher than 7 percent every year until they get net climate positive and start to regenerate natural capital.

So, what does the data say? Is genuine green growth really possible? Which corporations, cities, and countries are delivering healthy growth, and which are not? Before showing some select results, we need to take a hard look at that other fuzzy concept: "inclusive growth."

SETTING TARGETS FOR INCLUSIVE GROWTH

Measuring inclusive growth raises many questions: How can we know if we are heading to a more equitable society, or if our steps are large or miniscule? Are we dealing only with the climate crisis and not climate justice, too? Is inclusive growth measurable,[27] or is it just another feel-good buzzword from the la-la land of globalist politics and wishful thinking?

It may never be possible to set a uniform, measurable, science-based target on a socially optimal threshold of inequality, as we may never agree on a "right" level due to diversity in ethics, values, cultures, and worldviews. Even the United Nations views reducing inequality as a within-country issue, and hence does not set any globally defined numerical target for countries. But some approaches have nevertheless managed to formulate meaningful quantitative and consensus targets for inequality. Three of them are particularly useful.

Health-Based Threshold
As we saw in chapter 6, a large number of studies over recent decades have found strong evidence that rising inequality causes health problems, social conflict, and lack of education. That makes metrics applied to health indices applicable to broader inequality concerns. The studies on the inequality level above which health problems significantly worsen have so far converged on a Gini score of 0.30 as the threshold, which is comparable to a Palma ratio of around 1.1.[28]

Human Development and Happiness Thresholds
The United Nations' Human Development Index (HDI) ranks countries based on their citizens' life spans, education levels, and income levels. Its

World Happiness Report ranks countries based on surveys that assess how happy their citizens consider themselves to be. Both measures seem to correlate strongly to overall inequality. The top ten countries in the 2018 HDI, for instance, had an average Palma ratio of 1.1, while those in the "low human development" category had an average Palma ratio of 2.2.[29] Likewise, the top ten countries in the 2019 World Happiness Report also had a Palma of 1.1, while the bottom ten had a 3.0.[30]

Sustainable Development Goal 10

Unlike the data-based levels mentioned above, the United Nations' seventeen Sustainable Development Goals, described in chapter 6, are negotiated goals based on political consensus. The SDG 10 ("reduce inequality within and among countries") states: "By 2030, progressively achieve and sustain income growth of the bottom 40 per cent of the population at a rate higher than the national average."[31] This means that whatever share of national income a country's poorest 40 percent has, it should not decline any further in coming years. It is a goal for "shared prosperity," but it does not name a specific numerical target or rate of change. The richest 10 percent of the world already collects around 53 percent of the total income.[32] In light of this number, SDG 10 sets a relatively weak target for inequality: keeping the income share of the poorest at the current level, so that they don't lose out any more on their share of future growth to 2030. However feeble that may sound, there is a certain ambition to it because it means breaking the trend of worsening within-country inequality.

In 2014, Nobel Prize–winning economist Joseph Stiglitz and former UN assistant secretary-general Michael Doyle recommended a specific target to end the top-heavy income distributions that create extreme disparities and thwart sustainable development.[33] By 2030, they advised, all nations should have their richest 10 percent taking in no more income than the poorest 40 percent. That describes a Palma ratio of 1.0, and it is the level that is typical for the Nordic model. Sweden, Norway, Finland, and Denmark all have Palma ratios in the 0.9–1.0 range. They benefit from an "equality multiplier" that has left them not just more equitable and stable economically but also more efficient and flexible.[34] As explained in chapter 6, it is when the richest decile (top 10 percent) starts taking more than 40 percent of national income that they really squeeze the share of the poorest. This level of inequality both hinders growth and turns unhealthy for the whole population. The 1.0 Palma ratio has also been recommended by Oxfam, and in the SDG Index and dashboards.[35]

A SCIENCE-BASED COMPASS FOR THE EQUITABLE SOCIETY

It is reasonable to say, then, that the Palma ratio should be brought down toward 1 from the within-country global average of more than 2 to achieve a more stable society and healthy economy.[36] But many will object to setting normative recommendations based on empirical information. They would state that this is the realm of politics, not science—and they are fully right. Yet many policymakers prefer to lend an ear to data and science-based recommendations about impact and outcomes. Of course, an inequality metric cannot itself change national policies, but it may be a necessary basis for doing so. And the research seems to point unanimously to Palma ratios near 1 as being a stronger driver for social health, stability, *and* economic growth than those in the 2 to 7 range.

Given that economic growth since the 1980s has come with increased within-country inequality, growth and inequality need to be seen together. We need to measure the inclusiveness of growth by combining both its *rate* and its *distribution*. In order for growth to be sustained and effective in reducing inequality, it needs to benefit the poor more: the bottom 40 percent consistently need to get *larger* shares of the larger pie.

Thus, to reduce inequality in disposable incomes both within countries and globally in line with the three approaches above, I recommend *halving disposable income inequality by 2050 for any economy with high inequality today* (say, a Palma over 1.5).[37] Achieving this requires a greater than 2.3 percent reduction per year from 2020 to 2050. The pace at which income gaps narrow across countries will also depend on their economic growth. Therefore, at the same time, we need rapid economic growth (> 3 percent per year) in emerging-market and developing economies. If they keep that level, global inequality will get better over time because rich countries tend to grow more slowly (< 3 percent per year, sinking toward an average of < 1 percent per year closer to 2050 as their GDP per person goes above $40,000 in 2011$ purchasing power parity, PPP).

Just as resource productivity can measure annual progress toward genuine green growth, I propose the concept of social productivity (SOPRO) as an annual measure of fair, inclusive growth. While resource productivity is defined as the ratio of value creation/resource use, social productivity would be the ratio of value creation/inequality (as measured by the Palma ratio, which Stiglitz and Doyle also favor over Gini, since Palma is more intuitive and more easily communicates inequality levels to the public). The annual rates of change in social productivity (sp) must be substantially higher than the growth rate in order to be genuinely inclusive. In other words:

> SOPRO = social productivity of growth = value added / Palma ratio
>   of disposable incomes, in $
> sp = rate of change in SOPRO = Δ(GDP/Palma ratio) in % per year
> Inclusive growth is: $sp > g$

But how much higher must sp be than g to make countries sufficiently equitable? With an expected average economic growth of around 2.6 percent per year to 2050[38] and at least 2.3 percent reduction of inequality, we get a rule of thumb for how much this measure of inclusiveness must increase per year to move toward genuine fair growth: at 5 percent per year until the Palma ratio is near 1. Our guiding metric, then, for genuine fair growth becomes $sp > 5$ percent. This type of measurable inclusiveness, or social productivity of national income, is illustrated in figure 7.4. Inclusiveness is strongest in the shaded triangle that appears in the compass's top left (northwest) triangle labeled "Genuine Fair Growth." Achieving system change is all about shifting the course from the little black arrow that indicates average annual unfair growth since the 1980s around in the direction of that wide white arrow labeled "inclusiveness." This points toward truly inclusive (or pro-poor) growth in the sense that the economic pie is growing in size *and* the poor's share of that larger pie is increasing, too.

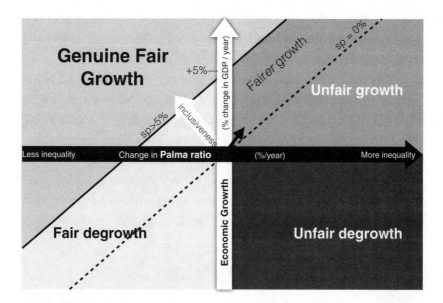

Figure 7.4

The genuine fair growth compass: The vertical axis shows the change in size of the economic pie. Larger is north (up), smaller is south (down). Moving east (right) on the horizontal axis means a larger share to the top 10 percent incomes. Moving west (left) means a larger share to the bottom 40 percent. Changing course to the northwest means a more inclusive growth, i.e. shifting from the black arrow (representing the global within-country average for 1985–2010) to the thick white arrow pointing northwest. That makes sure that the poorest 40 percent get a larger share of a bigger pie. Halving inequality while at least doubling the economy requires growth toward genuine fair growth (marked "$sp > 5\%$"). That can be achieved with at least 2.3 percent less inequality per year and average 2.6 percent economic growth, each year from 2020 to 2050.

If we repeat such inclusive growth every year from 2020 to 2050, with an average rate of change in social productivity greater than 5 percent per year, we can double national incomes while halving within-country income inequality.

Has this kind of fair growth recently been achieved by any country over longer periods of time? Comparing the United Kingdom, Sweden, United States, and Germany since 2000, the United Kingdom comes the closest, with an average of 3 percent annual gain in social productivity. Germany is now trending toward unfair growth with an average −1 percent *sp*, and in 2008–2009 it plunged into unfair degrowth when its *sp* was -8 percent due to the financial crisis hitting the poor disproportionately and stagnant incomes in low-skill jobs. Sweden's *sp* is averaging 1.9 percent since 2000, but falling short of the greater than 5 percent target has less severe consequences for Sweden, which is already an inclusive society with Palma ratio around 1. Sweden only needs to keep its Palma ratio stable while growing, which it has done from 2000 to 2016. The United States is doing better than Germany on *sp* but worse than Sweden or the UK. With an average annual *sp* of 0.9 percent, which is lower than its growth rate of 1.9 percent, it is still ensconced in the unfair growth quadrant (the US Palma increased from 2.6 in 2000 to 3.0 in 2014).[39]

CLEAR GUIDES FOR TRANSFORMATION

Solving the double crises of climate and inequality requires transformational rates of change in both the resource and social productivity of growth. But most societies change their deep socioeconomic structures slowly due to inertia. It's a decadal process. Yet, due to the nature of exponential growth, rates above 5 percent become transformational, giving rapid system change while rates below 5 percent are less perceptible and evolutionary. At 5 percent rates, the resource productivity will double every fifteen years. At 7 percent, it doubles each decade.

That is why compasses that show whether actual annual changes are *sufficient* to keep on track come in handy: They allow us to see early wins and annual steps, both over long periods of time and within the short-term cycles (two to four years) driven by investors and political time frames. They let us objectively discern the leaders from the laggards, year by year, blowing away the dual smoke screens of random or faraway targets, murky rankings, greenwashing, SDG washing, and pro-poor posturing.

To cross this vast ocean of transformation over three decades, we will need clear stars to steer by, helping to keep our course steady as public fears and fancies come and go. Our direction, after all, is mandatory: *Successful growth in the twenty-first century depends on shifting from maximizing economic growth alone to resource and social productivity growth.* And we can now define healthy growth clearly, succinctly, and objectively:

Healthy growth is $rp > 5$ percent and $sp > 5$ percent each year.

To realize such healthy growth we need individuals, investments, innovations, institutions, and international policies to keep the focus on both the direction and the speed, each year and every decade going forward to 2050.[40] We need clear signals on our own performance, social feedback, and visualizations on how fast change is happening for all players and in all parts of the economy. How are countries, cities, and companies actually measuring up so far in the twenty-first century?

CHECKING PROGRESS AT THE NATION LEVEL

Green growth critics frequently claim that there is no historic evidence of genuine green growth.[41] But that is wrong. The good news is that several countries have already demonstrated genuine green growth consistently over decades since they started aiming for it in the early 2000s. Among them are Sweden, Denmark, Finland, and the United Kingdom.

In figure 7.5, the more the curves turn upward, the more the economy is thriving while emissions fall rapidly.[42] The curves visualize the speed of systems change. From the figure, it's clear that Sweden, Denmark, and the United Kingdom have on average achieved the 5 percent carbon productivity per year threshold (see the "GGG" line) every year so far in this century. That's the average minimum capro required going forward to meet the 2°C climate target by 2050. Among the Nordic countries, only Norway lags, performing lower than the OECD average with regard to carbon productivity, primarily due to its increasing emissions from offshore oil and gas production.

Sadly, though, many of the largest economies of the world have *not* delivered genuine green growth since 2000. Their rates have improved however since the early 2000s, when the world average was only 1 percent per year. During the 2010s, all these heavyweights jumped their capro rates up to the 3–3.5 percent per year range (US 3.4 percent, EU 3.2 percent,

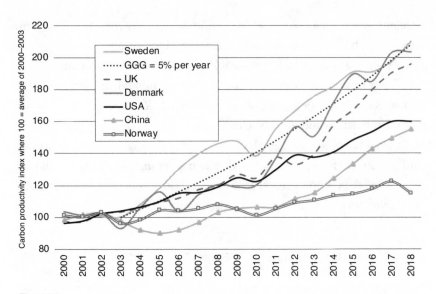

Figure 7.5

Who is doing their fair share of growing the economy with lower emissions? Genuine green growth requires carbon productivity (capro) to grow at higher rates than 5 percent per year, illustrated by the small-dotted line labeled "GGG." At that average rate of change, the global economy can thrive while emissions fall sufficiently to keep global warming below 2°C. The 100 mark represents the baseline at the start of the century—the country's average CAPRO of 2000–2003.

OECD 2.8 percent, BRIICS 3.3 percent per year).[43] This is quite hopeful, as it is then "only" yet another step of +2 percent per year that is needed for them to start coming on track to the 2°C pathway during the 2020s. That should be fully doable, owing to the four driving forces described in chapter 3 in this volume.

If the European Union (EU), for instance, manages to reach its current 2030 targets (-40 percent GHG emissions relative to 1990) while growing at its recent average rate of 1.8 percent GDP, it will achieve a capro of 4 percent per year. That's green growth, but not quite genuine green growth. However, the European Commission works at increasing the target to 55 percent cuts by 2030.[44] If they deliver on promise while growing GDP 1.8 percent annually to 2030, that's a capro of 6 percent per year—which is genuine green growth, even if still a little off from a true 1.5°C-compatible target. At the time of writing, the United States has no such goals.

How have Sweden, Denmark, Finland, and the United Kingdom managed a much higher capro than the EU, US, or OECD average? And are their performances replicable by other countries? Several studies have pointed to a number of drivers behind the development. For example, in recent decades these countries have accelerated a structural shift in jobs and value creation—moving from industry and manufacturing to service sectors that are less emission intensive, including information, communication, high-tech, and knowledge jobs.[45]

In addition, there are country-specific strategies: Sweden has since 2000 upheld a strong focus on energy efficiency and renewable energy, including the phaseout of oil for heating in the residential sector. A high and rising carbon tax on fossil fuels, first introduced in 1991 (in 2018 at 140 $/t$CO_2$, the highest in the world), turned biomass into the most competitive fuel for heating, which further decarbonized the country's fuel mix. Sweden also has an ambitious action plan to add more renewable capacity. It is also seeing strong growth in its digital tech and internet sector, with several innovative new high-growth companies like Spotify, Mojang, and iZettle. In all, Sweden is leading the way toward a low-carbon economy, reports the International Energy Agency (IEA). In 2019 it had the second-lowest CO_2 emissions per GDP among the member countries and the second-lowest CO_2 emissions per capita.[46]

Denmark has decarbonized its economy by using wind energy and natural gas instead of coal and oil, and increased energy efficiency through district heating and combined heat and power with bioenergy. Denmark also stimulated the growth of renewable energy and energy-efficiency industries by investing in R&D and smart cities and creating a domestic market for energy technologies, particularly in relation to wind. These new industries are estimated to add 1.6 percent to GDP and 1.5 percent to employment in Denmark. The stimulation of a domestic market for renewable energy is reflected in energy prices and, more specifically, in the relatively high energy tax burden for consumers.[47]

Despite being the laggard of the four Scandinavian countries, Norway generates almost all of its electricity from abundant and affordable hydropower, which is well integrated into the Scandinavian grid. Its use of hydroelectricity as its main energy carrier since the 1970s sets Norway apart from its neighbors and other countries.[48] The building sector is increasingly energy-efficient, and mainly runs on this hydroelectricity. Norway's power-intensive metallurgic industry has improved its carbon productivity substantially since 2000. Even so, this is countered by the operations of its offshore oil and gas industry—with many oil fields entering energy-demanding tail production stages, which generates greater emissions for each barrel produced. Emissions from the domestic transport sector have also increased since 2000, but have recently fallen due to rapid and widespread introduction of electric vehicles. The combination of emissions growth in both offshore and transport since 2000 cancels out other energy-efficiency measures, thereby thwarting Norway's ambitions for genuine green growth.[49]

The United Kingdom was lagging behind on genuine green growth in the first decade of the twenty-first century (with a capro of 2.5 percent per year). But it has shown sustained and historically rapid improvements in recent years, with a capro of 5.8 percent per year since 2010. Coal power has been phased out, and overall energy efficiency has improved, with wind and solar power gaining shares. Also, total gas consumption is down 30 percent in the UK compared to the early 2000s. Despite all the political noise around Brexit, the country has seen genuine green growth in recent years thanks to UK carbon law, carbon pricing, and improving efficiencies.

The United States has achieved some green growth since 2000 (capro 2.9 percent average per year from 2000–2018, with $g = 2\%$), but not genuine green growth. Emission reductions have come roughly one-third from the transition from coal to gas power. But equally one-third has come from energy efficiency and another third from renewables.[50]

The surprising star since 2010, however, is China. Now the largest economy in the world (based on purchasing power measures), China has achieved average levels of capro greater than 5 percent per year (2010–2017), as can be seen from its now rapidly rising trend line in figure 7.5. Back in 2000–2005, China had a *negative* capro each year—dirty growth. If the recent reported numbers are reasonably accurate, and if China can keep up this new record rate in the coming years and decades, we may witness a first sustained and sufficient rate of change in a major economy. China's *13th Five-Year Plan* for 2016–2021 set a 2020 carbon productivity target that is 50 percent above 2005 levels. It also sets a new 2030 target of 65 percent carbon productivity. This means China plans for a capro greater than 4.2 percent per year until 2030. Many are now expecting China, which achieved a capro of more than 7 percent per year during 2013–2016, to overachieve on these targets.[51] If the country can maintain such rates of change, that would certainly be promising.

What about inclusiveness? The United States' Palma ratio was at 3.0 in 2014—a high level of inequality.[52] From a total of $10 trillion in post-tax disposable incomes, the 10 percent richest took $3.9 trillion, three times the share of all the 40 percent poorest, who together took $1.3 trillion. The middle 50 percent took $4.8 trillion.

For countries with a Palma ratio of 2 or higher, evidence-based recommendations would call for halving inequality by 2050. This is more easily done if the economy simultaneously doubles, as measured by value added (which translates to +2 percent growth in GDP each year from 2015 to 2050). In such a scenario, everyone gets higher disposable incomes. Yet the poorest's share grows the fastest. The poor catch up substantially over the decades. This scenario of doubling an economy while halving its inequality is illustrated in table 7.1.[53]

By 2050, the slice going to the poorest grows from $1.3 trillion to $4.9 trillion (or from 13 percent of the total pie to 21 percent). The contrast between 2014 and 2050 is shown in figure 7.6.

Table 7.1

Historic values for the average income per adult in 2014, and a 2015–2050 scenario in which inequality is halved (reducing the Palma ratio from 3 to 1.5)

	Total net income trn$ (2014)	Income per adult (2014)	Total net income trn$ (2050)	Income per adult (2050)	% change 2020–2050	% change annually
Richest 10%	3.9	264,000	7.4	500,000	89%	1.8%
Mid-50%	4.8	64,000	11.0	153,000	138%	2.5%
Poorest 40%	1.3	22,000	4.9	83,000	276%	3.9%
Total	10.0		24.0			2.5%
Palma ratio	3.0		1.5		–50%	–2.0%

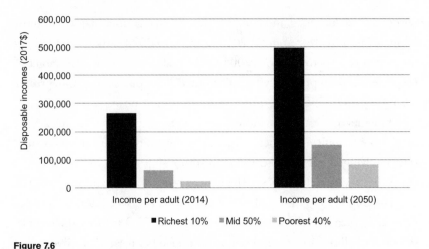

Figure 7.6
An inclusive growth scenario for the United States shows how inequality can be halved by 2050: The poorest 40 percent triple their disposable incomes over thirty-five years (2015–2050) while the mid-50 percent double theirs. The richest 10 percent grow their income with another 85 percent reaching an average of nearly $500,000 per person. Together, this halves inequality over thirty-five years, with a 2 percent decline in Palma ratio per year. Numbers in constant 2017 US dollars.

The central question when it comes to inclusive growth is: To what extent does economic growth benefit the poor? Inclusive growth, or pro-poor growth, means that the extra value creation in economic growth goes proportionally more to the poor. When this happens, social productivity rises rapidly ($sp > 5$ percent per year). Using this integrated metric, which combines value creation with inequality, we can now see what the recent historic performance looks like: Has growth recently eased disposable income inequality (after taxes and transfers) or made it worse? Figure 7.7 shows how well five countries are doing in letting economic growth benefit the poor, using tax data from 2000 to 2017.[54]

The more the curves turn upward, the better for the poor and hence for social capital. Each country starts from whatever inequality level it had in 2000. The graph doesn't indicate Palma disparities alone; it shows how well economic growth has been used to reduce inequality (up, as with the United Kingdom) or to worsen it (down, as with Germany). To solve the inequality crisis, countries with high inequality ought to rise steeper than the small-dotted curve every year.

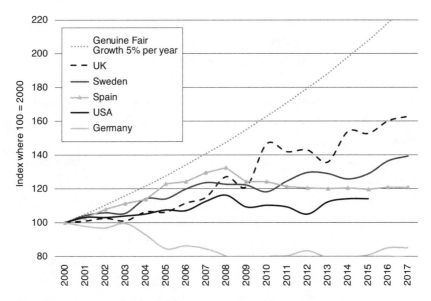

Figure 7.7

How has growth benefited the poor? The 5 percent genuine fair growth per year (small-dotted) line implies a transformational pathway for a country with high inequality (Palma ratio > 1.5). The lines below show the inclusive growth rates of five countries relative to their 2000 level. Fair growth is calculated as annual economic growth (GDP) divided by rate of change in inequality measured by the Palma ratio.

The United States started the 2000s with a high Palma ratio of 2.6. But tax data show it rose even further, to 3.0, by 2014, as most of the gains of stronger growth went to the richest 10 percent. Hence, the poorest 40 percent were hardly better off in 2014 than they were in 2000 (the curve only lifted from 100 to 114, and the median household income was flat). The benefits of the 2 percent annual economic growth went almost entirely to the 10 percent richest and, as we saw above, particularly to the 1 percent richest. Among this handful of countries, then, the United States had the second least inclusive growth, and remains very far from a transformational pathway.[55]

Of the countries shown, Germany has the most unequal development. It had slower growth (1.2 percent annual GDP growth in 2000 to 2016).[56] It also had rapidly worsening inequality of 2.2 percent per year, increasing its 2000 Palma ratio of 1.1 to 1.6 by 2016. Thus, it had negative inclusive growth—or, put plainly, a pro-rich, unfair growth—over the entire period. For the poorest 40 percent, there were no gains at all from the country's moderate economic growth.

The most inclusive growth among these countries was demonstrated by the United Kingdom, whose Palma ratio fell from 1.7 to 1.4 between 2000 and 2016. This was achieved through an annual economic growth of 1.8 percent, accompanied by a 1.2 percent reduction in inequality per year. Hence, during those years the country's annual average social productivity, or inclusiveness of growth, was 3 percent per year. The disposable incomes of the 40 percent poorest grew much more quickly (1.3 percent) than the top 10 percent (0.2 percent). This fascinating development merits more extensive analysis than there is room for here, but part of the explanation seems to be that the 2009 financial crisis hit the richest proportionally harder, paradoxically making the United Kingdom less unequal from 2009 to 2010, as can be seen from the little "hill" on the graph.[57]

In 2000, Sweden was already an equitable society with a Palma of 1.0, which it maintained all the way to 2017 while growing its economy. Thus, Sweden does not need to halve inequality further. The country only needs to continue to ensure that the share of the top 10 percent does not escalate higher than the share of the bottom 40 percent. With this steady rate of shared prosperity, both the poor and rich Swedes were on average 40 percent better off in 2017 than they were in 2000.

The simple conclusion from analyzing these countries' pathways is that none of the countries with higher Palmas (above 1.5) managed to come near the recommended transformational rate of $sp > 5$ percent. This is due to the plain fact that the political will to implement pro-poor policy alongside economic growth has been lacking, except in the Nordic countries. Like Sweden, Denmark, Norway, and Finland also have Palma ratios around 1, and hence don't need to aspire to a transformative inclusive pathway ($sp > 5$ percent) but only need to maintain their current level ($sp \geq g$).

This graph makes it clear that inequality doesn't solve itself through conventional growth and market dynamics alone, as neoliberal economists holding onto trickle-down ideals would have us believe. The countries with Palmas above 1.5 are not *at all* on track for doubling value creation while halving inequality by 2050.

What the *sp* > 5 percent metric can clearly show is the annual progress (or deterioration) of the otherwise fuzzy concept of inclusiveness or social justice. This could—in theory—make it easier to hold politicians accountable and build a public collective will to use more of the tools in the inequality toolbox. It also highlights the flaws in the dubious conviction that "a rising tide lifts all boats," as the examples of Germany and the United States show. Today, says Joseph Stiglitz, "the rising tide has only lifted the large yachts, and many of the smaller boats have been left dashed on the rocks."[58]

SCALING FROM THE COUNTRY TO THE COMPANY LEVEL

There are a vast number of metrics, key performance indicators, and rankings used to rate company performance along sustainability dimensions. There are so many, actually, it might be quite confusing to orient oneself among them, and many rankings lack transparency. The scores somehow just fall out of a black box that only the consultancy making them understands.[59]

But based on the principles from the previous chapters, I will provide a few examples and core arguments as to how and why companies could account for healthy growth in simpler, more transparent and comparable ways, based on numbers they already have (mostly) in their annual reporting.

To assess corporations' performance, it is essential to get a picture of their rate of change in resource productivity. This is done by connecting a company's value added (gross profits) to its resource use. The reason for choosing value added (rather than revenue or full-time employees, for instance) is that the value added from all economic entities, including companies, is by definition aggregated into a country's GDP. Just as we analyzed countries' resource productivity based on their GDP, we can analyze companies' resource productivity based on their value added as it scales well between the levels and across sectors. Again, we will start with climate and use the same capro indicator. A company's capro, too, needs to improve at more than 5 percent per year for it to do its fair share to decarbonize its value creation quickly enough to contribute to the 2°C target set by the Paris Agreement.

In several sustainability rankings, a few companies are consistently recognized as leading the pack. These often include the consumer goods company Unilever, the carpet manufacturer Interface, and the Danish energy company Orsted.[60] Walmart, too, is frequently mentioned, as are the energy company Equinor and global recycling-systems leader Tomra. How have these companies, perceived as green leaders in their class, actually performed over time based on the genuine green growth measures we have just defined? By gathering the value creation data for a handful of companies from public sources such as annual reports and financial databases (Factiva), and then dividing by greenhouse gas emissions (from CDP, scope 1+2),[61] we get the rates of change shown in figure 7.8.[62]

The small-dotted line delineating a 5 percent minimum improvement in capro is marked genuine green growth. For this time period, Interface, Orsted, Tomra, and Unilever have all succeeded in increasing value creation while reducing emissions sufficiently rapidly. If all other companies started copying their carbon productivity performance, humanity would have a better chance of solving the global heating. Interface, for instance, has reached a capro of 20 percent in recent years, and Orsted has reached around 15 percent.

The companies below that 5 percent line continue to emit a lot of greenhouse gas emissions per dollar of value creation. Walmart, for instance, is pushing its suppliers hard to cut their emissions but is failing to do enough with its own (scope 1+2). And Equinor—often mentioned as an environmentally leading petroleum company[63]—did not improve its carbon productivity at all since 2010, despite massive PR campaigns portraying its oil as "cleaner" than other companies' oil.

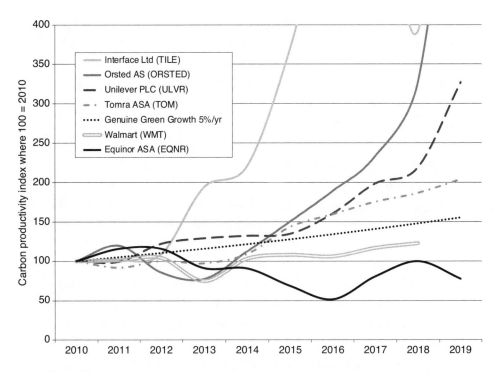

Figure 7.8

Carbon productivity of companies widely considered leaders. A company's carbon productivity is its value creation in constant currency divided by tons of greenhouse gas emissions from its own operations and energy use. Each company's rate of change (capro) is shown relative to its earlier performance, where 2010 = 100. The small-dotted line illustrating genuine green growth shows an annual 5 percent improvement, a science-based target, which is the minimum global average for transformation needed to stay below 2°C warming.

SCOPING, CARBON FOOTPRINTS, AND WHAT MATTERS MOST: PRODUCTION OR CONSUMPTION?

Critics of green growth are eager to point out that certain aspects of carbon footprint assessments can lead to false success stories—both at the company and country level. They ask how nations like Sweden, Denmark, the United Kingdom, the United States, and Germany can be touted as emissions-reduction leaders when they achieve their higher carbon productivity mostly by outsourcing their dirty production to emerging economies where environmental standards are lower—sometimes making their real footprint even higher than if they had kept their dirty production at home.[64] They wonder, for example, how a construction company that purchases cement can avoid adding those production emissions to its own footprint when making its concrete products.

Indeed, one could argue that a country's footprint should include the embodied energy and emissions from the production of all its imports. By calculating emissions from home-based production and from imports and exports you would get a so-called consumption-based footprint.[65] The same argument can be made at the corporate level: A company's emissions reporting should include emissions from all of its purchases as well as the impacts of its products at customers' locations.

Both arguments seem logical, but when you dive into the details, questions emerge. Where do you draw the boundary between which impacts companies or countries are held accountable for, and which they are not? Who is responsible for improving a product's carbon and resource productivity at each point in its value chain and life cycle?[66]

The answers lie in *scoping*—a key tool for assessing emissions and, as such, also a key tool for using the carbon or resource productivity metrics introduced in this book to determine whether growth is grey or green. But those answers are not as straight-forward as they might seem. To solve it, we need to use the so-called scope 1, 2, and 3 with the growth compass. The leading standard for quantifying and managing carbon footprints is the Greenhouse Gas Protocol, a system aimed at making emissions assessment and reporting practices uniform. It is used worldwide to define three scopes for accounting carbon footprints.[67]

- *Scope 1: All direct emissions* from activities fully under the control of an organization or country. Examples include all fossil fuel combustion, industrial process emissions, and refrigerant leaks at a company's own operations, or within a country's territory.
- *Scope 2: Indirect emissions* are the emissions from electricity purchased and used in production. These emissions come from power plants outside of scope 1, owned by someone else—typically a utility.

- *Scope 3: All other indirect emissions* from suppliers and customers, occurring at sites and sources that the reporting entity does not own or control. Examples include business travel, procurement, and waste- and water-treatment facilities.

The three scopes set up a very clear moral prioritization. An entity has primary responsibility for its scope 1 emissions, secondary responsibility for scope 2, and tertiary responsibility to improve scope 3 emissions from one year to the next at a sufficient rate of change. The entity can't use improvements in scope 3 impacts as an excuse for doing nothing to emissions in scope 1 or 2.

So, when calculating the footprint of a product, company, city, or country to measure progress with the growth compass, we can and should focus on reducing scope 1 emissions first. When reports on scope 1 shows genuine green growth (capro > 5 percent), or if scope 1 emissions are already near zero (as for a consulting company that only rents office space, having no direct emissions), then it's time to report on scope 1 + 2 and improve that. Finally, all companies and countries could report on 1 + 2 + 3 emissions.

The carbon productivity metric really comes in three versions: $capro_1$ for scope 1, $capro_{1+2}$ for scope 1 + 2, and $capro_{1+2+3}$ for scope 1 + 2 + 3. For the examples in figure 7.8, I used $capro_{1+2}$.

Based on concerns about countries and companies offshoring or outsourcing their pollution, it would seem that scope 3 should play a central role in emissions assessments. However, there are two main arguments against overemphasizing scope 3 emissions accounting:

- All human-caused emissions come from somebody's scope 1. In other words, all actors have primary responsibility for the emissions occurring within their own territory or otherwise under their direct control. When applying the sufficient-green-growth metric presented in ch. 5, these actors are also responsible for securing capro > 5 percent for their scope 1. When this is increasingly applied globally, it will eventually secure the sufficient rates of change toward green growth throughout the world economy.

- Scope 1 reporting is spreading everywhere; all countries report to the UNFCCC, and companies are increasingly reporting to CDP. As they do, any attempt to include scope 3 emissions performance from other countries, customers, or suppliers can result in duplicate accounting. The same emissions (and reductions) will be accounted for both in importing and exporting countries, and both at the customer and supplier sides of corporate trade.

Here is where questions about boundaries get murky. If you buy grain, do you include emissions generated from soil or forest loss, or at the farm? And what about companies that are tempted—in their annual reporting of carbon productivity—to include their products' positive effects on their customers' emissions? They may even be tempted to exclude some negative impacts from their own supply chain. For example, a company that primarily uses hydropower to make aluminum and has a much smaller carbon footprint than its coal-based competitors may claim that exporting cleaner aluminum creates avoided emissions that should be counted to the company's benefit. Should it really get a credit from scope 3 that it can use to offset its own scope 1 emissions?

The simplest way out of this accounting conundrum is to always start with rapidly improving domestic carbon productivity for countries, and capro (scope 1 and 1+2) for corporations. National policymakers can then bring more and more entities across world trade in line with capro > 5 percent standards, for instance with border tax adjustments based on embodied carbon in imports.[68] This process would accelerate consistent carbon accounting. And corporations can step up green procurement requirements everywhere, see chapter 8 in this volume, to include more and more of the entire life cycle footprints of their global purchases (by reporting also on $capro_{1+2+3}$). Finally, the value chain of all products would eventually be covered.

HEALTHY GROWTH IN CITIES AND SECTORS

More and more, cities are moving into the foreground of climate action. They have become hotbeds of creativity and innovation that compete for

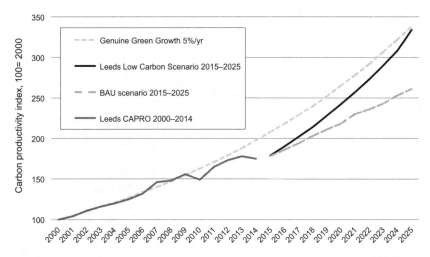

Figure 7.9
Carbon productivity for Leeds, United Kingdom, (calculated using gross value added in USD per person 2011 PPP/GHG emissions). The index 100 = the year 2000.

global talent and capital. With sufficient investment, they can lead the way in turning billions of people's lives around to low-carbon lifestyles and a green, circular economy.[69]

One city whose performance has been well mapped by researchers is Leeds in the United Kingdom. The Leeds City Region has a population of over three million and an economy worth over US$86.2 billion, approximately 5 percent of the UK economy. Its genuine green growth performance is illustrated in figure 7.9.[70] For the 2000–2014 period its CAPRO seemed to do very well, as shown by the solid line rising along with the required genuine green growth rates (small-dotted line). But after the financial crisis, performance started lagging up to 2015. What will happen to 2025? Researchers explored the economic case for climate action in Leeds going forward.

The two lines for 2015–2025 show one business-as-usual (dashed line) scenario and one low-carbon scenario (solid, black line). In the latter, at no additional cost and with a payback time of seven years on low-carbon investments, Leeds manages to catch up and rejoin the genuine green growth pathway by 2025. Leeds could achieve this by investing in the

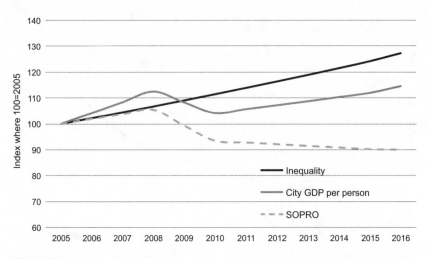

Figure 7.10

Inclusive growth in Leeds? City GDP (or gross value added, in USD constant 2011 PPP) for the Leeds region went up almost 20 percent from 2005 to 2016. But inequality also went up 22 percent—which implies that inclusiveness (or social productivity) declined. The poor have received little or none of the benefits from growth, i.e. unfair growth over the period.

usual suspects: insulation of buildings, new renewable energy, heat pumps, biomass heating boilers, and reduced household heating levels.[71]

When it comes to inclusiveness, however, Leeds has seen increasing levels of inequality. Its Palma ratio is up from 1.1 in 2005 to 1.4 by 2016.[72] When combined with city GDP per person, we can determine the inclusiveness of the growth as seen in figure 7.10.

Any city can now monitor whether it is doing its fair share to locally confront global challenges like climate change and inequality. The metrics remain the same: Genuine green growth requires annual *rp* improvements of at least 5 percent, and genuine fair growth requires annual *sp* improvements of at least 5 percent. A city using these metrics can also start benchmarking its rates of change with neighboring cities, triggering a positive competition that can accelerate change. In addition, the visualization of the social inclusiveness of growth, based on the Palma ratio (disposable income from tax data), makes it easy to see whether growth is used to benefit the poorest (curve goes up) or only the lucky few (curve stays flat or declines).

The same principles can be applied in accounting for different economic sectors as well. Buildings, energy, agriculture, forestry, industry, and transportation all have vastly different characteristics. Yet to come together in green growth in the entire economy, all sectors need to be analyzed in a comparable way (check the rates of change in resource productivity to see if $rp > 5$ percent). Otherwise wasteful consumption may partly rebound and just shift from one sector to another. The good news is that abundant innovations are available today—in all sectors—to improve resource productivity by a factor of five or higher in the coming decades.[73] But again, without proper accounting and comparison it's impossible to see if each sector is doing at least its minimum fair share.

AT THE HEART OF HEALTHY GROWTH: CREATING VALUE

Core to healthy growth is the value creation process—delivering something that other people want and value in money. But whether you create value added by purveying wooden bowls, renovating a building, teaching a class, instructing yoga, or designing software, the magic lies in meeting and exchanging with someone who appreciates your contribution enough to pay you. These interchanges can build trust and relationships, dynamics that form the core of human well-being and happiness and foster the true social value added that free markets, free exchange, and free trade can create.

Market capitalism can be an unruly beast, though. We all know how markets can be alienating and sickening, with mechanized, impersonal transactions, power abuse, corruption, intolerable inequality and fraud. They can leave both customer and workers disempowered and feeling screwed. They can rip up forests and ravage wildlife populations as the ecosystems around them are fouled. These are cases of badly designed, badly regulated, or badly governed markets.

Yet, since markets are the basis of our cities and our civilizations, I prefer to try to work on improving and redesigning them rather than turning against them altogether. And to do that we need an (at least) three-dimensional growth compass, especially as we go deeper into the twenty-first century, that can visualize how value creation—the heart of markets—is changing and nudge it in a healthier direction.

First, value creation sets the stage for profitable, economically thriving ventures, cities, or countries. It generates jobs, wealth and taxes. It's also the basis for calculating the labor and capital productivity (the value of output per hour and per unit of capital). With more value creation (measured as gross profits in the accounts), a company can reinvest and grow more, driving innovative change. Second, resource productivity lets us monitor the carbon, water, nutrients, biocapacity, and other resources consumed in our enterprises and reduce the resource use per unit of value added until the total environmental footprint fits safely within ecosystem and planetary boundaries. We can then move beyond net zero extracted resource use and actually reinvest in nature, becoming regenerative. Like trees. Finally, distribution patterns let us see whether the value creation is shared among workers, shareholders, finance, and government in such a way that inequality is kept within reasonable bounds. The objective is not at all to make everyone equal but to avoid rampant inequality (Palma > 1.5), which tends to make societies sicker and more violent.

The three main productivity measures—labor, resource, and social productivity—are mutually reinforcing *for each and every dollar made.* When all three aspects of value creation are connected in the three-dimensional healthy growth compass, the diagonal move from one extreme corner (unfair, gray degrowth) to the other (fair, green growth) is one along which human well-being increases. Human flourishing is best served when all three dimensions are interconnected. In contrast to conventional growth, healthy growth is defined by a balanced growth in broad assets. It is the opposite of single-minded growth, in which one type of capital gains at the expense of others. It allows us to grow laterally into an understanding of wealth that is more complex and cyclical than what GDP and other financial metrics alone can describe or measure. This broader, balanced view of capital that includes natural and social capitals, was missing from the mainstream economics and accounting that dominated the entire twentieth century. It was also missing from policy. Through the lens of price only, Solow and others failed to realize that when focus narrows to just financial and human capital (hours worked × pay) along with technology, our broad wealth itself suffers.

With long-term healthy growth, each generation can pass on to the next a better broad wealth than it inherited.[74] One form of capital should not be allowed to eat up all the others. To get a healthy economy, at least two of the three must increase without the third declining. This principle is both a core conservative, progressive and environmental value, and hence should be able to gather broad nonpartisan support. But if countries fail to look after one or more of their capitals properly over time, then the next generation will be worse off, even if the GDP has grown.

III HOW? GETTING PRACTICAL

A BRIEF INTRODUCTION TO THE TRIANGULAR SYSTEM

Conventional economic growth is stuck in a catch-22. It has made our Earth's ecosystems sick, disrupting 7,000-year-old ecological norms.[1] Yet, to flourish both poor and modern civilizations need growing economies that rely on those ecosystems being well and safe.

Healthy growth, then, has two missions: First heal the faults inside conventional growth, and then turn further growth into a healing remedy for the Earth's life-supporting systems. In this way healthy growth can start undoing the damage that unfair, gray growth has already done to both the social fabric and ecosystems. Second, it can move up to the next level—becoming regenerative of natural and social capital. This is the basic principle of psychotherapy: take a seemingly pathological symptom and help turn it into a healing remedy. Can this work not only on suffering individuals, but also on a grand systemic level?

Many thinkers who want radical system change call for an overhaul of its central feature: the marketplace and its capitalist core engine. The idea is that the interchange between market and consumer, ruled by big corporations and the finance system, is too broken to fix. But healthy growth can happen within the market, too, if we allow ourselves to understand the market—and its key players—differently.

In most standard economic textbooks, you'll find the classic depiction of how our economy works: the circular flow diagram (figure IIIa). It has come to define our view of "the market," simplified into households (consumers) on one side, businesses (firms) on the other, and circular

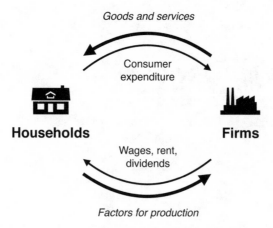

Figure IIIa

The economy as the circular flows—"the market"—between households and companies/firms.

arrows looping them together. The point is to show how money flows from households to companies each time we buy goods or services, and how money flows from firms to households each payday and when owners get other income like rent payments or investment dividends.

What this influential circular flow diagram leaves out, however, is a third group of key players: governments, including agencies, laws, and courts. In any economy there are companies, households, and the government that supports the two. These three make up an interdependent societal system, which scaffolds our everyday economic lives. And around and in between all three lies nature—our living habitat and provider of beauty, water, air, foodstuffs, and more (figure IIIb).

For our economies to accelerate healthy growth, all three players must change in sync. If consumers are to buy goods and services that are both resource and socially productive, those goods must be affordable and easily available. If companies are to sell more of the productive goods and services, there must be enough consumers with a willingness to pay. For companies to have a level playing field in which to compete, they need equal access to infrastructure, an educated workforce and a just legal system, among other things. These must be provided by government. For government to provide these things "for free" to companies, it needs revenues and taxes to

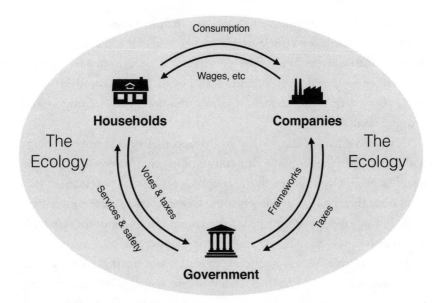

Figure IIIb

Society as triangular system inside the ecology: showing interactions among the three main groups of households, companies, and government.

be generated by the companies. For citizens to improve their health, freedom and mutual trust, they need schools, city administrations and police. To provide these things to everyone, taxes on extraction, consumption and incomes are needed.

Thus, we have an economic system where no progress occurs if only one group changes. Say a business offers up an innovative solar power system with storage and blockchain peer-to-peer trade functionality. But there are neither private nor public buyers nor a well-functioning grid. The product line, however helpful to society, will fold quickly. Or imagine there are lots of households who want to buy zero-emission electric cars. But there are no affordable models with good range on the market, nor enough public chargers with standard plugs. Consumers will end up buying something more accessible and convenient, such as a conventional gas guzzler. And imagine if there are solar panel pushers that sell cheap, poorly made products that break down in a few years. Or electric car providers that use batteries that degrade much more quickly than promised. Or too many charging standards for vehicle-to-grid integration. Or too many dubious financial instruments. Then active governance is needed, not to "strangle the market" with red tape, as many fear, but to provide coordination of processes and frameworks for optimizing the design of the market and product standards.

This three-part synergy is key to markets that benefit buyers, sellers, and society over the long run. The conventional circular flow diagram was deeply flawed. Healthy growth thus depends on better market design that recognizes three, not two, cornerstones. Governments secure people freedom and equal opportunity, guard the commons, and enforce standards and transparency. They have a coordinating, entrepreneurial, stimulating, and guardian role. Yet in order to see that role, and use it as a lever in system transformation, we have to replace the neoliberal minimalist framing of government as incompetent, bloated, and strangling. Better governance does not necessarily mean bigger government (measured in the number of pages of regulations, laws, or public servants). Reinventing governance must happen alongside the reinvention of products, technologies, and markets, within ecological boundaries.

The point is: The three groups have to act in sync to get the transformation going. But are they in sync? Misalignment creates excesses and inertia and the endless frustrations that many complain about. That's where the pain often lies, as when regulations (of financial products, for example) come too late, or there is a supply of organic foods but too little consumer demand, or an urgent ecological need for carbon pricing but no citizen support. Or when most citizens await climate leadership from politicians while politicians await public support for carbon pricing or other ambitious measures—but mostly hear outspoken opposition to them. It can be a vicious circle where next to nothing gets done.

But therein lies the opportunity to shift a system dominated by a self-reinforcing inertia. The same feedbacks, managed differently, can result in rapid change, particularly during crises or extreme events. The vicious circle becomes a virtuous circle in which one move by households is quickly accommodated by business and fine-tuned by government to optimize the overall benefits for households, which again benefits both government and business.

The triangular system doesn't generate just private gain and public pain, or vice versa. Rather, it builds on the private gain to strengthen the broad wealth of the commons so that the commons can support private gain, and the circle continues on. In part III, where we look at *how* to catalyze the type and rate of change we need, we'll explore this smarter system redesign. The next three chapters will explain how companies, individuals, and government can transform in unison to support healthy growth. What can you do, now, wherever you operate in this triangular system?

8 HOW TO REORIENT COMPANIES TOWARD HEALTHY GROWTH

The company, wrote two editors of the *Economist*, is the "most important organization in the world . . . the basis for prosperity of the West and the best hope for the future of the rest of the world." Over recent centuries, they noted, the rising power of corporations and the mounting consequences of their propensity to do both good and evil came to define human lives as much as—and then more than—other prevailing forces like governments or religion.[1]

Indeed, as critical as it is to restore the power of people and governments in shaping the future, right now corporations seem to have the upper hand. Views about whether they will innovate us to a better world or drown us in a sea of spoils are just as divided as are views on growth and degrowth. But there's a growing awareness among executives that companies need to change themselves before the world can change, and a worldwide movement has coalesced to drive that paradigm shift.

One of this movement's many, many leaders is John Montgomery, a Silicon Valley attorney who came to question why even good people overseeing corporations so often leave their values behind when entering the boardroom. Having worked for decades with public and private companies and in venture capital financings, he saw firsthand how it had become, as he says, morally acceptable "to maximize profit for shareholders and foist all the negative consequences of corporate behavior that you can get away with off on the commons."[2]

How did society come to permit such a destructive form of shareholder supremacy? In a conversation over a long lunch in Berkeley, Montgomery traces that answer all the way back to the founding fathers, who, he says, never addressed a fundamental philosophical question: "If corporations are persons, what rights and what moral responsibilities do they have?" At the time they were writing the Constitution, there were only about fifty corporations in America, and all of them were smallish, fringe things. Articulating the limits of their rights and the scope of their responsibilities was not top of mind. And so that legal framework has been shaped by a series of Supreme Court decisions that span nearly 200 years and continually strengthen the legal standing of companies, putting them on par—in many circumstances—with the rights of people. The controversial Citizens United decision in 2010 brought debates about corporate personhood to the fore.

Global corporations have revenues, assets, political interests, and thus powers larger than many nation-states. They can employ a much larger army of top-notch lawyers to fight their cause than any individual can. Add to that a sole purpose to maximize financial gains, and "the market" has created a monster capable of devouring both equality and ecosystems, society and soil, people and planet. Or the perfect psychopath: willing to use any and all means, including the courts, to override geographical and national borders, social concerns, and our natural commons and fight unceasingly for maximum short-term, self-interested gain.[3]

Simply running companies on solar power—however good—doesn't make them regenerative or redistributive. And if management, for some seemingly idealistic reason, does prioritize social concerns over shareholders, sometimes the shareholders can sue them or oust do-gooders from senior positions. Just the threat of such action is enough to discourage some executives. It then becomes a strong social norm among managers to "stick to core business" and "maximize shareholder returns." As a result, they guide their companies to shove costs onto others, avoid investments in long-term innovations, overwork employees and use profits for buy-back of shares to the benefit of owners and C-suite bonuses. They employ more creativity in tax dodging than in innovation.

Changing such norms has become Montgomery's life work. In 2010 he was among a group of pioneers pushing for California to legalize a new business structure, the benefit corporation, an entity whose legally defined goals include operating transparently and in the interest of its workers, its larger community, and society and the environment in general—not just its shareholders.

Rather than trying to upend 200 years of tilted Supreme Court decisions, and a deeply ingrained set of social norms among business professionals, the idea was to make a game-changing fix to the fundamental legal architecture of corporations through a simple modification in their articles of association. Any corporation that wanted to avoid the risk of becoming a complete sociopath could heal itself by adopting benefit corporation status. The legislation passed in California, as similar legislation had in five states beforehand.

THE RISE OF THE BENEFIT CORPORATION

After a company becomes a benefit corporation, it can change its status from a limited liability company (LLC) to a public benefit corporation (PBC). It then moves forward with its business—except now it has an explicit, broader purpose that both encourages and requires its directors to seek smarter ways to create multiple benefits, for employees, communities, ecosystems, and equity owners. In other words, PBC status gives legal teeth to the triple bottom line of "people, planet, and profits," the phrase coined by sustainability expert John Elkington back in the 1990s.[4] (The benefit corporation is a legal structure, while a *B Corp* is a voluntary certification process, building on the triple bottom line, available to any company.)

Today, more than thirty-six US states have adopted PBC legislation. Most important among them is Delaware, home to the majority of the Fortune 1000 thanks to its specialized business courts and traditional corporate statutes that assign just one legitimate purpose: to maximize stockholder welfare.[5] The number of benefit corporations rose from around 2,000 in 2015 to more than 5,000 in 2018. And it is not just small companies that are choosing PCB incorporation. This legal overhaul is yet another accelerating innovation, part of the sixth wave, that will keep picking up momentum.

As with anything involving human development and social systems, though, the momentum won't grow by itself. We need to keep encouraging and inspiring companies to choose this path, and also keep researching the performance of PBCs.[6] In that way, the smart, healing growth can gradually outperform those conventional corporations that stick to maximizing short-term, gray growth.

There are more than enough unconverted corporations to nudge for a long time. But they don't all have to convert to PBCs. Arguments can be made that LLCs also have the capability to turn around to include broader purposes and capitals without changing their legal structure—particularly since evidence is building that such healthy growth is more profitable over the long term. Hence, good governance for social and environment performance are already for the stockholders' benefit too.

A host of thought leaders tell us that companies can be critical change-makers. "Business is the only mechanism on the planet today powerful enough to produce the changes necessary to reverse global environmental and social degradation," says author and eco-innovator Paul Hawken.[7] Leading British environmentalist Jonathon Porritt says business leaders are "more predisposed" than politicians to drive sustainable development for the foreseeable future.[8] Several studies have found that most CEOs already report that they see sustainability as critical to their future success. Yet they're honest enough to also report they struggle with how to integrate healthy sustainability into their core strategy.[9] The shift doesn't happen by itself. It requires green executives, as management writer Gareth Kane calls them.[10] And large investors are increasing their demands and expectations for so-called ESG (environment, social, governance) reporting and performance.[11]

"The crucial thing is cultivating leadership from the heart," stresses Montgomery. He suggests three guiding steps: (1) business concerns expand to include social and environmental consciences; (2) companies then measure what matters to reflect these; and (3) companies design jobs that work for human beings. The good news, based on leading companies' experiences, is that any business in any sector can transform into green, profitable growth. It's possible to work bottom-up, progressing from more cosmetic outer actions to changes that affect the core business at the same time as top-down. And new businesses can establish healthy business models right from the start, then keep improving. It's a two-way stairs—walking the talk up and down in parallel.

The ability to manage with sustainability in mind is becoming a key competency for leaders, which is why I teach it at the Norwegian Business School in Oslo. I've taught hundreds of executive-level students how to use the following tool to start transforming their companies.[12]

SIX STEPS ANY COMPANY CAN TAKE TOWARD HEALTHY GROWTH

If business is to be transformational, where do executives and employees start? And if leadership starts coming from the heart as well as the wallet, what do the hands start doing Monday morning? This chapter introduces six practical, tested, strategic steps that any organization, public or private, LLC or PBC, can use to profit from the shift to resource productivity, clean energy, and social co-benefits. The steps progress from changes in what I call the outer level (outside the business' daily boundaries) all the way into the inner level (changes in the internal operating system). Somewhere along the way up the stairs a company could choose to change its articles of association to a benefit corporation. At the top, a company has aligned its core business with the healing growth model.

The six strategic steps are:

1. *Outreach*: Taking actions that benefit people or the environment but are external to the organization, such as collaborative research and development projects; giving philanthropic support to NGOs, community projects, or churches; engaging in lobbying efforts; improving industry standards; or engaging in voluntary carbon offset schemes.

2. *Housekeeping*: Cleaning up one's own buildings, workspaces, and assets with environmental and labor management systems that continually improve energy efficiency, safety, compliance, waste handling, regulatory compliance, and governance. A company on this step might

install rooftop solar, and improve race and gender balance at executive levels and worker representation on boards.

3. *Supply:* Setting sustainability requirements for all suppliers; buying fewer but better and longer-lived materials; ensuring fair trade throughout the supply chain. This forces resource productivity—and social responsibility—from the whole supply chain by leveraging the power of procurement.

4. *Operations:* Redesigning operations for optimal resource productivity through whole system redesign. In many cases, factor four improvements or better can be found, relative to conventional industrial levels.

5. *Product portfolio*: Phasing out products that are wasteful in production, in use, or after use, and cannibalizing their market shares with new and innovative products or services.

6. *Business models*: Shifting the business model toward a dematerialized or circular way of creating more value for all customers and stakeholders, including employees and shareholders. This is the final and most important step.

The six steps can be illustrated with the "stairs model" of healthy growth strategies shown in figure 8.1.

Some companies start at the bottom and move up. Others—for instance, a new solar energy startup—may start at the top with a green-to-the-core business model, then later progress outward to the housecleaning of their offices and outreach. For most conventional organizations, though, transitioning to a healthy growth model means addressing issues at all six levels.

The lower three or four steps are focused on being responsible and reducing negative impacts. They are mainly about doing "less bad." The higher steps focus on scaling the positive opportunities of healthy growth: new profitable offerings that make more value for customers by solving their problems in leaner, smarter ways with a net-positive footprint and social benefits. They are about doing "more good," healing the impacted ecosystems, workers, and communities.

Since more complex and deeply internal changes always come with risk for unexpected costs, detours, and delays, it may be smart to start with

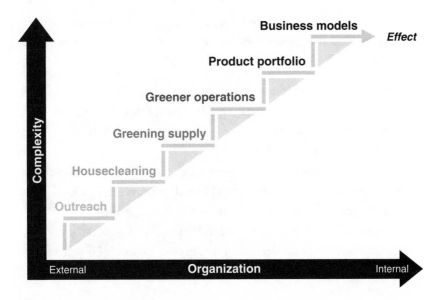

Figure 8.1

The six steps model for healthy growth, usable by any organization that wants to systematically identify opportunities for transformation. One can start with early wins, and then progress up toward larger effects on profits, ecology, and society. The vertical axis shows that these steps start with simple measures, and then grow in complexity. The horizontal axis shows that each step moves the transformation inward, toward the core of the organizational DNA.

the easy, early wins at lower levels. But moving up the stairs brings ever larger impacts and better financial results. In addition, research shows that when a company starts implementing healthy growth initiatives, employees thrive more at work, improve their productivity, and develop more loyalty to the company.

Outreach

Sometimes outreach is chided as "corporate cosmetics" and dismissed as window dressing. Yet it's fundamental. From a systems point of view, a company is embedded in a set of relationships. Not just with customers and owners, but also with knowledge clusters, city centers, research institutes, neighborhoods, parks, ecosystems, rivers, local and national NGOs, police, and many more stakeholders. Each organization must conduct a proper review to find out who its stakeholders are, get into dialogues, and prioritize them according to mutual impact and human needs.

The software company Salesforce, which provides customer relationship management for corporate clients, has an extensive outreach program it calls the 1-1-1 integrated philanthropy model:

- 1 percent of their product subscriptions are given for free (or at a deep discount) to nonprofit clients. The important thing was to define how their product could help solve a social problem through the nonprofit's mission.
- 1 percent of employee time is given away. Salesforce sets up volunteer programs that include team or individual volunteering, with recognition programs for volunteers, and outcome measurement.
- 1 percent of equity is put in a philanthropy fund in a local community foundation. The foundation and its directors establish how these dollars can be used into the community, completely independent of the current CEO.

Since 2000, Salesforce.org has delivered more than $300 million in grants, provided more than 4.3 million employee volunteer hours, and powered 44,000 nonprofits with free Salesforce technology.[13] More than 700 other companies have adopted the 1-1-1 model through Pledge 1%, a corporate philanthropy movement dedicated to making the community a key stakeholder in every business. Salesforce CEO Marc Benioff, instrumental in launching Pledge 1%, famously stated, "The business of business is improving the state of the world."

Countries like Norway and Sweden have long traditions of large companies and cooperatives devoting significant portions of their assets to supporting sports, culture, and the arts. The SpareBank foundation, for instance, holds more than 25 percent of the bank's shares and uses proceeds to support the local communities in which it operates. The Gjensidige Foundation owns 62 percent of the publicly listed insurance company Gjensidige. The foundation gives most of its dividends back to its customers, the insured, but also gives away around 20 percent of its yearly dividends to local projects geared toward health and safety.

Important as corporate philanthropy is, participating in joint strategic research projects or broad industry collaboration initiatives can turn out to have even more wide-ranging impacts. In the early internet days, startup technology companies wanted information to be free, and made open source software (such as the Netscape browser, which evolved into Firefox) available to everyone. Tesla conducted generous outreach when it opened up its patents. By inviting every car manufacturer to join the electric car industry, Tesla hastened electrification of the entire transport industry. Many large companies, too, have joined the Breakthrough Energy Coalition, an initiative fronted by Microsoft founder Bill Gates that puts private and public money together into shared research and development efforts that can accelerate the energy transition.

And just after Trump's election in 2016, nearly 400 companies and nonprofits signed a letter in support of the Paris climate agreement.[14] When businesses raise their voices alongside scientists, the message travels farther inside policy circles (though not in all administrations).

The outreach step offers low-hanging fruit to busy leaders who want to start moving toward green and fair leadership. It doesn't disturb internal operations. It doesn't upset sales efforts or demand internal restructuring. People outside the company do most of the work. Outreach activities can contribute to making the company more profitable since they strengthen stakeholder relations and reduce risks to the company's reputation. That lowered reputation risk is one of the four main drivers of sustainable profitability, according to a research review conducted by a team at Oxford University and Arabesque, a financial think tank.[15]

With outreach, you can start fixing old negative external impacts— doing good for others while also doing well—even if such contributions come at a short-term cost.

Housekeeping

Whether securing the handling of or eliminating toxic materials, ensuring the health and safety of employees, improving workspaces, stopping unnecessary waste, conserving clean water, or improving energy efficiency, good housekeeping is about reducing and recycling waste as well as fixing the leaking pipes and valves. It's about serving tastier, healthier foods at the company restaurant, and generating less food waste and wrapping while doing so. Or encouraging biking or taking public transport to work.

Good housekeeping entails what initially may seem like a lot of nitty-gritty environmental and human resource management.[16] Target, measure, implement, evaluate. Celebrate early wins and milestones. Then repeat. And, as soon as a company grows to a certain size, it needs a proper management system to do it well. The ISO 14001 system is the leading international standard for environmental management systems. There are long checklists in such systems that may feel overly bureaucratic. But housekeeping also entails dialogues with people in other parts of the company about what can be better, smarter, cleaner, and safer. It is about designing jobs with dignity. If well conducted, housekeeping efforts spur enthusiasm for more interaction across organizational silos, including on how to set more aligned performance targets, as well as more collaboration on learning and on how to get continually better together.

The upside potential is that—when well done—housekeeping improves overall company performance, reduces risks, and improves employee motivation and morale. Researchers at the University of California, Los Angeles and Paris Dauphine University surveyed around 5,000 people from 2,000 companies that were onboarding environment management systems. They found that the knowledge sharing and learning involved in making these improvements boosted employee motivation and commitment to the organization. They also found that businesses following international standards (like ISO 14001 environmental, organic, or fair trade standards) see a 16 percent increase in labor productivity compared to companies with lower standards.[17]

Central to good housekeeping is improved waste management. The famous waste hierarchy says we should always reduce material consumption first, then reuse as much as possible, and finally recycle. And if nothing else works—let it go to energy recovery (burning) and only then, reluctantly, to a landfill. It's also important to take this a step further: Redesign the processes that create waste on the company's premises to prevent as much as possible, preferably eliminating it altogether in a zero-waste scenario.[18] But integrating this action in a company-wide, comprehensive way typically happens farther along in the stairs model—in step 3 (supply), 4 (operations), and 5 (product portfolio).

Most housekeeping work is about moving toward zero impacts: zero waste to landfill, zero spills, zero leaks, zero use of toxic materials, zero accidents, and so on. Ironically, petroleum companies that are used to toxic, hazardous gases and explosive chemicals often have excellent health, safety, and environment (HSE) systems in place. They go out of their way to make sure that each ladder, valve, garage, trash can, and engine is operated safely, according to standards. Equinor—the Norwegian oil giant—even has separate safety regulations on how employees back up their cars in their parking lots to avoid accidents. Climate writer George Marshall was once stopped at Shell facilities by an employee who pointed out that his shoelace was loose. He was told that tying his shoelace and holding the handrail can reduce accidents.[19] The irony, of course, is that by understanding environment in this narrow sense (only their own workspace), Shell creates a frame inside which they believe they've delivered on health, safety, and environment requirements—yet all the while their main product destabilizes the entire climate, threatening all humans, slowly but certainly bringing devastation everywhere.

In short, comprehensive housekeeping steps are essential to have in place before leaping to more complex changes. After putting your own house in order, it is possible—with a moral legitimacy—to require all your suppliers to do the same. That's why the next step involves leveraging the power of procurement.

Supply

Because both private and public organizations buy an incredible amount of stuff from each other, top procurement officers have vast amounts of money running through their hands. What if they decided to gradually turn more and more of those sums toward ever more resource-productive, better-quality goods rather than always going for the cheapest initial price? How would their actions ripple through the national and international trade that is the lifeblood of our economic system? Many companies exert most of their impact on nature early in their supply chains. When companies and governments start to leverage the power of procurement, huge shifts can happen downstream.

One unlikely green champion in this regard is Walmart. With more than 100,000 suppliers worldwide, Walmart's initiative to require better sustainability standards from all their suppliers has the potential for vast impact. In 2017, Walmart launched Project Gigaton, to eliminate one billion metric tons of greenhouse (a gigaton, or 2 percent of global emissions) gases from their global value chain by 2030. To qualify as a Walmart supplier, a company has to answer fifteen detailed questions about whether, and by how much, they are reducing energy costs and greenhouse gas emissions and improving material efficiency by reducing waste and enhancing the quality. Suppliers are also asked to report on the human and community issues where they operate. Do they know all the manufacturing locations where their products are made? And do they invest in community development activities in these locations?

Certainly, most suppliers have a long way to go to give Walmart the answers it wants. But the good news is Walmart has discovered that there is so much resource waste in its supply chains that squeezing that waste out can be good both for sustainability and for saving money. With leaner and more productive supply, Walmart can probably sell greener products at the same or even lower price. And even if Walmart for years have not paid its own employees decently,[20] they want their suppliers to treat their employees well so that no scandal hits the Walmart brand.

IKEA, another behemoth, has achieved more sustainable sourcing for all of its cotton and the wood used for its products.[21] The company directly collaborates with and invests in its suppliers, recently allocating over $3 billion for sustainability investments in its supply chain. This includes securing a long-term supply of wood materials by investing in sustainable forestry, as well as in suppliers that are active in recycling, renewable energy, and biomaterial developments.

Rather than each company developing its own requirements (and generating loads of reporting work for suppliers), several standards are being developed. For instance, the electronics company Philips has joined the Responsible Business Alliance (RBA) and built on that for its own procurement staff. The Philips Supplier Sustainability Declaration covers labor and human rights, worker health and safety, environmental impact, ethics, and management systems. In line with Philips's own general business principles, it includes the right of freedom of association and collective bargaining. The declaration also requires suppliers to cascade the RBA Code—as a total supply chain initiative—down to their next-tier suppliers.[22]

If Philips's procurement team notices a delay in reporting or implementation of a corrective action plan, several levels of consequences may be triggered. In exceptional cases where a supplier is unwilling to improve, Philips terminates the business relationship. However, to prevent audit fatigue and limit the burden of audit preparation and follow-up at a single supplier site, Philips and other members of the Responsible Business Alliance have agreed to share audit results. This eliminates the need for multiple audits and enables a stronger focus on corrective actions and their follow-up.

By pushing best-practice requirements down to each supplier, these big companies may have larger effects than national regulations can have. And by using the cascade principle, their standards may eventually reach all the way down the supply chain to mining, energy, and agriculture.[23] There will be glitches, corruption, and fraud, of course, as the shadow side of both humans and corporations is always present. But the threat of losing the opportunity to sell to these major companies is enough to give most suppliers a strong nudge toward compliance—stronger than many national regulations, which may be weakly enforced in many countries.

Procurement teams can add value by influencing suppliers' operations, their product portfolio, and their future product design. They can also contact their own internal operations managers, nudging them to think about a more long-term procurement strategy at an earlier stage. Then they can help look for smart solutions. The idea is to think in systems and functions rather than just specifications and cheapest initial cost. Smart procurement officers may then have time to source innovative solutions from better suppliers so that the company can purchase less but better stuff that lasts longer. And the after-use of the goods may even already be designed through cradle-to-cradle principles not just for recycling but for upcycling.[24] The real benefits only become apparent when there is time and opportunity to conduct life-cycle analyses and life-cycle costing. Today, unfortunately, these are still underutilized tools, also relative to cost-benefit analyses. Overlooking them imposes extra long-term costs both for the company and for nature, and the wastefulness continues.

Another key success criterion for greening procurement is that the procurement officers are required to report to senior management about their progress. This ensures that the procurement people who stick their necks out for innovative, green, durable, and responsible new purchases are acknowledged and rewarded for their efforts.

Operations

Continuing up the stairs, we advance to the fourth level. Here, our full attention is turned from suppliers to improvements in internal production processes and service delivery. Incremental improvements in industrial production might, for starters, involve replacing one's old boilers, heaters, heat pumps, fans, insulation, or converters with the best available technology and newest products on the market. It can be shifting from fossils to renewable energy. Changes like these can yield substantial gains in resource productivity and profits over time, not least by reducing operating expenses.

But the fun (and the substantial profits) start to take off when one envisions more innovative redesigns. By aiming higher one can sometimes improve energy efficiency by over 90 percent.[25] Such redesign may include a whole system approach to rethinking a company's manufacturing processes and technologies from scratch, or even replacing physical resources with electrons and information systems.

One highly profiled renewables initiative is the RE100. Launched by the Climate Group and CDP, which runs a global disclosure system for environmental impacts, RE100 promotes the benefits of companies going 100 percent renewable, with their own operations run by self-owned renewable power sources.[26] Both Google and Apple, among the two largest companies in the world by market cap, are investing heavily in sun and wind to run 100 percent of their estate and power-hungry server parks. This greens their operations by making them less carbon-intensive, but it doesn't really redesign or improve them. They still have still large, centralized power-hungry server parks, only they now munch in-house wind or solar power rather than coal power purchased from the old utility company. IKEA, too, is investing heavily in its own power production, with massive rooftop solar panels on its stores. In 2020 IKEA, too, hit its target of making as much renewable energy as it consumes in its operations.[27]

Two Norwegian examples of operational improvement are Finnfjord and Norcem. Finnfjord is one of the world's largest producers of ferrosilicon, a metal used in steel manufacture and recycling. It discovered that it could recapture waste heat from the energy-intensive metallurgic process, regenerating enough steam to cover 40 percent of total production needs. The investment paid for itself in four to five years. Norcem is the largest Norwegian producer of cement, which causes 2 percent of Norway's greenhouse gas emissions. Worldwide, the production of cement for concrete is an even larger emitter, around 5–8 percent of global emissions (twice the amount of all the world's aviation). Norcem's ambition is to produce cement with zero emissions from production by 2030. It has improved energy efficiency by reusing process heat. It replaces coal as heating fuel by burning waste that otherwise would end up in the landfill or incinerator. And they mix fly ash, otherwise a waste product, into the cement in order to improve it. Altogether, this cuts emissions by 40 percent for each ton of cement, profitably. Norcem also plans to put a carbon capture and storage unit on its chimneys to extract final CO_2 from the limestone calcination process, which could bring future emissions from operations to near zero. Since cured concrete itself absorbs some CO_2 from the air over time, the cement may finally become net-positive.

The above examples are mostly incremental improvements that are important and relatively low risk. Yet there are limits to how much can be achieved by this piecemeal efficiency approach. The whole-system approach to resource productivity, however, allows for even greater benefits. Not just a "better boiler," but a new design of the energy system. For instance, one of the world's largest and most well-known industrial facilities, Tesla's battery Gigafactory, replaced the proposed gas boilers of 1000 kilowatt thermal capacity with better designed heat pumps for solvent redistillation, at 15 kilowatt electrical capacity. With that, Tesla achieved a stupendous

98.5 percent on-site energy saving, reports Rocky Mountain Institute's visionary Amory Lovins.[28]

The French-Norwegian company TiZir runs a metallurgical plant on a stunningly beautiful fjord on the west coast of Norway. They make titanium oxide—a common pigment in all white paints and colors, as well as iron—using vast amounts of the mineral ilmenite, coal, and hydropower. Being run on renewable hydropower, their production was somewhat climate friendlier than their competitors', but they still used a lot of coal, accounting for close to 1 percent of Norway's emissions. They wanted to expand. But CEO Harald Grande, a local from that fjord, and his team were concerned about climate emissions and local air quality. Wanting to get rid of coal completely, they developed a whole new production system, in which they can use hydrogen rather than coal to reduce ilmenite. The hydrogen can be made from water with the abundant, cheap hydropower from the area. Then, when hydrogen takes the oxygen out of ilmenite, resulting in the valuable titanium products, the only off-gas is water. Thus TiZir can triple its production, use 40 percent less energy, and emit 90 percent less CO_2 from the entire process.[29] These new hydrogen-system reactors are currently being tested and scaled. If and when they succeed at scale, the system can also be transferred to steel production (which accounts for another 5–8 percent of global emissions).[30]

Product Portfolio

Not long ago, I received my brand-new, blue, electric BMW i3, replacing the old Prius hybrid I got six years earlier, when I got rid of my old diesel Nissan SUV. The BMW i3 gets 120 miles per gallon equivalent, compared to 35 mpg for the Nissan. Now I relish in the thought of never having to fuel up with gas or diesel again. This little sequence of cars illustrates green growth along the product portfolio: from diesel SUV to fuel-efficient hybrid to electric car run on renewable power. Car manufacturers will still make and sell cars, but fewer noisy, exhaust-making ones and more and more electric ones with better sourced and circular materials.

The housekeeping, supply, and operations steps are often about reducing the negatives. This product portfolio level is where the real fun begins: changing all products and services to resource-productive, durable ones; using only ecosystem-compatible, nontoxic materials; maximizing renewable energy and sources; and applying cradle-to-cradle thinking to design for bringing benefits before use, in use, after use, and in reuse.

Gradually, the set of products or services one sells is shifted toward better (for the customer), greener (for the ecosystem), and more profitable (for the employees and shareholders) offerings. In the process, a company may even cannibalize the market share of its older products in the process. Eventually, more and more products will get a net-positive and then regenerative impact. Examples abound, and new innovations include AlterEco chocolates, Patagonia's Long Root Beer, Pukka Herb Teas, Cocopallet a pallet made from waste coconut husks, Qpinch heat recovery, Hunton Nativo wood fiber insulation, natural leather Piñatex made from pineapple leaves and many more.

Conventional executives and designers often assume that achieving such radical resource productivity is too expensive. But Amory Lovins has done it repeatedly with 10xE, the Factor Ten Engineering initiative, demonstrating how to make integrative product design into a repeatable discipline. And Cradle to Cradle Products Innovation Institute, founded by architect William McDonough and chemist Michael Braungart, provides guidance, principles, and certification for making products circular on a regular basis. These and other initiatives demonstrate that very large—up to 90 percent—energy and resource savings can be very profitable across a wide range of applications.[31] Sensors and artificial intelligence can be added to every appliance, truck, and building for huge energy, water, and material savings. Already, washing machines can adjust the amount of water by the weight of laundry and how dirty it is, houses can adjust lighting to match daylight and residents' comings and goings, and trucks can improve their loading and driving by cloud-based logistics. 3-D printing can supply local on-demand solutions and key components for better maintenance and durability of products. Thus, by replacing stuff with information, a megatrend often referred to as digitalization, companies contribute to the dematerialization of the economy. That's the key to making more value from products and services that consume ever fewer new materials.

The product portfolio step builds on three main principles: Start replacing all existing products and services with more resource-productive, digital, and eco-efficient ones. Use only nontoxic materials while maximizing renewable energy. And apply cradle-to-cradle thinking to design for bringing benefits before use, in use, reuse, and after use.

Business Models

A business model explains in a few words or figures how value is created in a unique way, and for whom, by the company's activities. The typical business model of the 1900s industrial age manufacturer was: grab—make—sell—trash. The value created was made by minimizing the cost of the grabbing (resource extraction) and the making (labor productivity), while maximizing the revenues and having the product trashed as soon as possible so that consumers would come back for more.

This has of course already started to change. But many manufacturers and brands remain stuck in twentieth-century thinking. However, some of the world's most valuable companies (like Google, Facebook, Airbnb, or Uber) no longer sell any "stuff." Among the most valuable business models we now find free search for everyone (Google), something that was unthinkable just twenty years ago. There are also free social connections (Facebook, WeChat, Twitter, Tinder). The service provider Airbnb is positioned to become more valuable than the largest hotel chains that own hundreds of expensive physical hotels.

Why buy your own car and pay for maintaining, fueling, and parking it when you can get mobility as a service anytime you need it? This can happen today simply by booking a Lyft on your phone or in the future by grabbing a ride in a self-driving electric car.

China's Broad Group has evolved from being an manufacturer of air-conditioners to offering complementary services as a building energy service company and manufacturer of sustainable prefabricated buildings. It has expanded its main business model with additional profit streams, making the overall business more sustainable.

Some companies already have—and were founded with—green business models. Tomra was founded in the 1970s to help shops recycle beverage bottles. They developed a simple reverse vending machine to automate the task and a smarter sensor to recognize the bottles, calculate a value, and give the customer the right deposit back. They have now morphed into a company that uses sensor-based solutions for optimal resource productivity in a swath of industries. This means simply that they want to realize their vision of leading the resource revolution in the growing circular economy. They now deliver sorting solutions for metals, plastics, and foodstuffs such as nuts and potatoes, and help mining companies sort low-grade ore from commercial grades. Tomra makes more money by selling efficient solutions that reduce the waste in any material flow where smart, automated collection or sorting is needed.

Kebony, another Norwegian company, was founded in the 1990s to transform ordinary pine wood into hardwood that's very similar in look and quality to teak and mahogany. This is done by steam-boiling pine planks with a special alcohol developed from organic waste products. The treated timber has a different cellular structure that makes it look just like teak and is extremely resistant to rot and fungi. Kebony's business model is grounded in killing off illegal rainforest hardwood logging by giving customers a better, sustainable option made from responsibly harvested temperate or boreal forests. The wood is completely nontoxic and durable. The more Kebony produces and sells, the less need there is for tropical hardwood logging or chemically infused rot-resistant wood.

WANTED: LEADERSHIP

The global consultancy Accenture surveyed 1,000 CEOs from twenty-seven industries across 103 countries, to find out how important sustainability issues are to the future success of their companies. A whopping 97 percent of these executives said it is either very important or important.[32]

So why aren't already-enlightened businesses doing even more? Why are there still so many businesses that haven't even started climbing the stairs yet?

For most executives, the main barrier is the challenge of integrating healthy growth practices into core business functions.[33] A lack of available company-internal capital also holds them back from investing in innovative and company-wide approaches like the six steps above. But, crucially, it remains difficult for them to quantify the commercial benefits in a convincing way. When they crunch the numbers, the chief financial officers often aren't impressed. They see better profitability, with possibly shorter payback time (return on investment) in other potential projects. They believe the company's limited investment funds can give higher, short-term returns or stock-price hikes through competing projects such as better branding campaigns, entering new markets, acquiring smaller competitors, stock buybacks and the like.

Thus sustainability continues to be "pigeonholed as a marginal issue," writes Accenture's analysts, "still regarded by many companies as an extra cost to be cut in the face of short-term financial pressures, rather than as a core part of strategies to generate value through revenue growth, cost reduction or brand impact."[34] Some CEOs see it as outside their main strategy and their core competence.

This is where real leadership comes in. Many existing companies have cultures, strategies, and accounting systems where a healthy growth philosophy hasn't yet been integrated. Despite mouthing support, in practice the attitudes have been negative. They equate "green" and "sustainability" with granola eaters, do-gooders, climate activists, and higher costs, and they can't envision themselves within that frame. These mental barriers stop them from reviewing green, fair practices and spending time making the business case for them. Thus, existing organization systems and mental models make it difficult to see the future financial benefits of transforming up the stairs. Also, healthy growth is a long-term issue, not usually something that shows immediate results on the next quarterly earnings report or two.

Yet there are by now thousands of examples, studies, reviews, and meta-reviews concluding that sustainability has a positive impact across all industries. In these studies, five typical factors stand out: (1) it's the right thing to do; (2) it yields better performance; (3) it improves brand and reputation; (4) it aids risk management; and (5) it helps retain and recruit talent.[35] But green leadership is needed to explicate the business case and then implement it. *Climbing the stairs doesn't happen by itself.* It requires vision and the integration of healthy growth opportunities into core strategy, leadership, and management all the way up.

That translates into three main leadership ingredients: words, money, action. The *words* must be there to give direction. A leader might not only use those words in speeches to employees and media but also work to get them into a company's articles of association, becoming a benefit corporation, as John Montgomery advocates. In that way directors and shareholders are brought on board, too. Without *money* to back them up, the targets and the words will just be empty greenwash from management and boards. There must be real investment in innovations at the housekeeping, operations, product portfolio, and new business models levels to create more value. The business case must be soundly calculated for each step. Then, *action* is required to execute the strategy: implementing it, following up on it, accounting for it, reporting about it, and scaling it.

The ambition level is crucial too. Should a leader launch moonshot missions with breakthrough targets and big hairy audacious goals? Or incremental improvements? Some say we need 10X impact, not 10 percent. And quickly. Such moonshot missions can have much more motivational and charismatic effect on engagement. And to a large extent, they are correct: it's far too late for the 1–2 percent per year, gradual, smallish steps, which could have solved it if business had got started with them back in the 1980s. And yes, we need also trailblazers, like Interface, Unilever, and Orsted, who break the speed limit in their transformation and lift the average. Yet speaking grandly about audacious goals and moon-landings in ten years' time can also take the attention away from starting the journey today. It is a case of both-and, not either-or: to *both* deliver on at least 5 percent resource and social productivity per year now, *and* to aim for far larger innovative breakthroughs. Many executives are appropriately worried about too much bragging and green-hyping. A key value among industry professionals is to not overpromise and underdeliver—but rather to underpromise and then overdeliver. The first breaks down trust. The latter builds credibility, trustworthiness, and long-term value.

Finally, many CEOs—with a busy schedule driven by their core business and the latest crises—hesitate to jump on the express train and express a cautious, weary frustration that sustainability measures are moving so slowly in their markets as well.[36] Neither government nor customers seem to be pushing hard for it. There is no proper price on carbon emissions, and little consumer willingness to pay more for durable, greener products. Investors have been wary. So executives choose a wait-and-see attitude. But that's the opposite of the combined purpose-led and strategic leadership needed to get things moving. If they wait too long, they risk that start-ups or competitors will harvest the benefits. The sixth wave is rising.

Yet, it's equally true that the broad transformation of most businesses would happen more quickly if support from customers and government was stronger. Some executives feel they have taken their company as far as they can. Now the consumer market and government have to get going as well, if the virtuous circle in the triangular system is to pick up speed.

9 HOW CAN I DO ANYTHING THAT MAKES A DIFFERENCE?

One crisp afternoon between Christmas and New Year's, my wife and I were walking up to the mountains behind the small farm where I lived more than a decade ago. We passed a grove of trees where I used to walk my now gone and much missed Belgian shepherd. I could almost see him running between the old pines there like he used to do. Farther up, I looked back to see the trees standing on a ridge, as if on the spine of an enormous whale. Maybe a hundred of them together, in a carpet of crusted-over snow. There was, as always, a sense of perfect proportion and distribution among them.

Ten years ago, I discovered a few large, freshly overturned stones on the outskirts of the grove, upturned by an excavator when our neighbor who owns this land was clearing a logging road up there. Not long after, during a neighborly chat, I brought up the pine grove. I told him that we used to walk to it often, and that I was very fond of it. Now, if you're from the city, telling your third-generation farmer neighbor what to do—and not to do—on his land is never a good idea. Yet somehow I found the courage to politely ask him not to log that grove.

Today, the grove is still standing, despite the logging road being completed all the way up to the overturned stones. Some pines farther down were indeed logged, but up here the sun's evening rays still flowed between pine boughs as they always have. I like to believe that my speaking up for the trees had some impact. You never know. Maybe he wouldn't have logged anyway. Maybe they will be logged next year. Either way, I imagine there is a certain gratitude from the trees when they "see" me, similar to what I feel when I see them. One individual action, one hundred trees saved.

When thinking about all the world's problems, it is easy to slip into the helplessness trap. The more you think about the unyielding forces we are up against, the more a certain mental pattern—learned helplessness—takes hold. Learned helplessness, which has been researched extensively in psychology, is a natural response to repeated, overwhelmingly repressive situations. Thus, many have trained themselves subconsciously to believe they have no control over the situation, and so they don't even try again. Burnout and depression lurk just under the surface of many alarmed or concerned citizens after years of pushing hope and activism. When nothing you do as an isolated individual is capable of making a real difference, apathy or cynicism seem to be the only sensible options: it's tempting to just withdraw for the storm to pass. Maybe life will still be livable after it does.

Yet that rarely works out well, either. Apathy, sarcasm, and cynicism hardly take you on the pathway of well-being and flourishing.

Countering learned helplessness requires us to challenge the perception that whatever we do doesn't matter. Despite the repeated experience of previous past defeats, this does not determine our current situation. Nor our future. Life is an inherently open, unpredictable unfolding. We have to balance two truths: Despite trying over and over, voluntary action by individuals will never be enough to stop or reverse global heating or rein our economic activities back into ecological and social boundaries, *by itself.* These large problems are structural, and they need structural—not just individual—action. But everyone can, by making their own small or bold actions clearly visible to others, influence social norms that ripple up to build support for ambitious action at the business, city, state, and national levels.[1] These ambitious actions can—and eventually will—force the necessary changes.

Most of us play at least four basic roles that open up distinct scopes of influence on the triangular system: as consumers, workers, owners, and citizens.

AS CONSUMERS: VOTING WITH YOUR WALLET NEVER GETS OLD

We've heard for decades that leveraging our individual and shared purchases as customers can have a huge impact on corporate leaders, if we let them know through words, inquiries, and money that we do indeed care about the impact of their offerings. Despite not being a silver bullet, it remains as true as ever, as figure 9.1 shows.[2]

When customers shift their purchases, companies follow. This is consumer power. But how far are consumers actually willing to use that power in prioritizing products marketed as sustainable? The consumer research company Nielsen found that in 2018, the share of consumers saying they would definitely or probably change their consumption habits to reduce

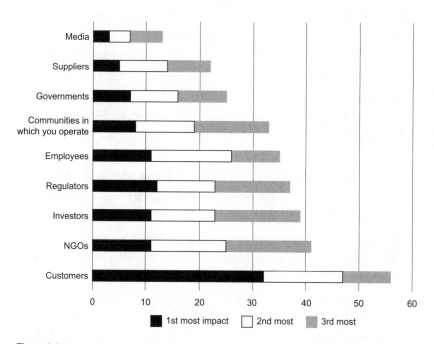

Figure 9.1

Who influences CEOs and corporations the most? Responses from 200 US multinational companies on the question: "Please select and rank the three stakeholder groups who you believe have the most impact on how you approach sustainability in your business."

their impact on the environment had grown across the world to a huge majority (73 percent). In the US, however, that number was 48 percent on average, but the younger generation (the millennials) were above global average (at 75 percent). Half of global respondents (49 percent) said they're even inclined to pay higher-than-average prices for products with high quality/safety standards, which consumers often associate with strong sustainability practices.[3]

Still, the market share of sustainable products (products that claim to be clean, no artificial ingredients, organic, sustainable, etc.) has in reality been rather low—at around 15 percent of the total US consumer goods market. The good news is that during the 2010s this market share has clearly grown more quickly. The annual dollar share of sustainability-marketed products grew from 14 percent to 17 percent from 2013 to 2018. That may not seem much, but it means that sustainable product sales grew by nearly 20 percent over those years. Specifically, in 2018 there was 5.8 percent growth rate for those sustainability benefits compared with 0.4 percent of conventional products. In other words, sustainable products are now where the growth is, in an otherwise stagnant market. Estimates to 2023 show 23 percent sales growth in sustainable consumer packaged goods, up to $140 billion.[4]

These are the kind of customer trends companies follow closely (see figure 9.1). But one has to wonder if the number of supposedly green and fair products that have proliferated have helped or hurt overall *material* consumption levels. How tempting is it for the person used to washing silverware after a large party to decide to order compostable forks instead, now that they are available on Amazon? People are, after all, hunters and gatherers at heart. We love to procure. So, if and when we spend our dollars on consumer goods, we should aim for those with a far smaller environmental footprint, and eventually a net-positive footprint as the ultimate goal. The whole consumer goods sector needs to turn around with sufficient rates of change in resource productivity, too (chapter 7). To accelerate this further, more of consumers' money needs to be shifted in that direction.

But why, when a clear majority share (50–75 percent) of customers say they want to buy sustainable, do sustainable products still retain a much lower market share (14–17 percent)? How can this untapped potential be realized? Psychologists have studied this attitude–behavior gap intensely and have arrived at several models to explain why some people behave in environmentally and socially responsible ways while others don't.[5] The findings point to the interplay of three main factors: motivation, capacity, and situation.[6]

Motivation grows from a person's own unique guiding values. All humans hold egoistic, altruistic, and wildlife values. But, from person to person, each value holds different weight and blends with different socially influenced norms and attitudes—which in turn shape our *intention to act*. Let's say I hold a deep love for animals, including cows and foxes. Then I might set an intention to avoid buying conventional milk, beef, and furry clothes and decide to spend my money on vegetarian food or conservation initiatives instead. But my family or friends may feel criticized if I go completely vegan: They serve and eat meat. This may weaken my motivation and cause me to delay or diminish my intended actions.

Capacity consists of knowledge, habits, and resources. Once I hold an intention, I'll need knowledge to act upon it. I'll find myself paying attention to information that I might have ignored before it became relevant to my intention. My rationale may expand as I learn, for instance, the water and climate footprint of a pound of beef. Then there are habits. Each of us has a personal history of habits—a set of behavior patterns that are easily triggered, run on autopilot, and may be fiercely resistant to change.[7] Finally, capacity includes my resources. If my intention is to bike to work more often, then I'll need to get a good enough bike, at least with brakes and no flat tires. My capacity relies on having the resources that can somehow match my intention to act.

Situation encompasses the external structural possibilities: that mouthwatering plant-based foods are available when I'm hungry, that it's possible to bike the distance to work, that when driving electric cars for longer distances there are chargers available on the route. When situations are conducive to intentions, we might still need cues to trigger positive behavior. If my intention is to go for more hikes during the week, or join a town hall meeting, or a community garden session, then a reminder from my smartphone might cue me into turning that intention to act into an actual behavior. Otherwise I might forget.

There are always barriers against taking action. I may be unsure about the environmental impact of the products I'm buying. There's too much information to review. I don't have the time to consider all the alternatives. Isn't this label criticized for greenwashing? I just can't make up my mind about what to do. Let's postpone it.[8]

However, if motivation and capacity are sufficiently strong, then I'll probably take action in spite of higher price, social disapproval, inconvenience, or other limitations. Learning about these factors and how they interact—both personally and on a theoretical level—may help us in sustaining individual action that can heal habits that are self-destructive, on a personal and planetary scale. But we still need nudging, something the booming fields of behavioral economics and economic psychology are all over right now. Nudging is the science of how to design the capacity and situation factors in such a way that it becomes easy and convenient to make the "right" decisions.[9] However, psychology reminds us that situation and capacity are not all. Values also matter hugely.

Rather than pontificating over consumption itself as evil, as some activists tend to do, we can remember that the way we let money flow, whether we have little or much, is a way in which to express who we are. I don't believe in the transformative power of asking people to become less materialistic and sacrifice buying what they really yearn for. What we consumers in modern market democracies need, however, is probably to become *deeper materialists*, who really care for the physical matter we buy and already own. We could maybe even have a much higher-quality consumption (measured in money) while consuming much less stuff (measured in tons). Fewer, but more beautiful and durable things. In any case, by redirecting our consumption spending to better align with our values, we can reward businesses that truly contribute to healing growth and create huge potential ripple effects. We can use our power as consumers to change the signals that determine the future direction of business and markets. If businesses listen to anything at all, it's to where your money flows.

AS WORKERS: ACTIVISM IMPROVES EMPLOYEE MOTIVATION

Whatever position we might hold in a company or organization, we can always inspire and influence our colleagues, teams, or bosses to take certain steps. That's not to say it's easy. Employees face many barriers in larger organizations with tons of procedures, budgets, short-term objectives, silos, routines, controls, and deadlines. Suggestions or pleas for more investment in green or social improvements are regularly brushed aside by overworked middle managers.[10] Good ideas get shot down daily due to internal company politics and power struggles. And employees shy away from even voicing them when cost cutting or internal competition create implicit or explicit threats of being fired or downsized. It can feel like circling sharks are just waiting to take a bite from your soft underbelly if you expose it in an unguarded moment.

Yet many managers and executives are in fact responsive to good ideas from employees, particularly if presented in a way that aligns with company strategies and emphasizes the positive outcomes. Half of all executives consider their employees second only to consumers in driving their approach to sustainability in the coming years. They rank higher than government, communities, and media.[11]

Workers, alone or in concert, can be a sort of corporate conscience. The crucial work of unions is of course to reduce inequality by fighting for fair pay so that labor's share of income does not sink any farther relative to the owner's share, and starts to climb again (see ch. 6 this volume). But unions have been challenged to expand their concerns beyond compensation and safety and take on more environmental and social issues. Some initiatives have resulted, but there is much room to grow.[12] Workers in cooperatives or (partly) employee-owned companies have a unique opportunity, too, to suggest and support initiatives that can secure and improve profits by shifting from gray to healthy growth strategies.

How to start? Push for getting promising opportunities up the six steps described in chapter 7 integrated into company strategy and execution.

AS OWNERS: LET YOUR ASSETS WORK FOR THE WORLD

Bonds, corporate stock, real estate: These are the three main assets that wealth managers, in their infinite wisdom, suggest you should have in a balanced portfolio. Particularly if you have more than a million dollars to invest. If you don't, they probably won't waste their time talking to you. But that's okay, because you can still do what is important: make your assets work for you and the future at the same time.

Let's start with real estate. Whether you own an apartment, a house, or commercial real estate, there are loads of opportunities to invest in those improvements we've known for decades we should make: better insulation, lighting, windows, heating, cooling. Every year, new technologies and products with better resource productivity are coming out. Windows, for instance, can now be tailored to let more daylight in, shade heat out, generate power during the day, and even change characteristics dynamically according to your needs.[13] Call a zero-upfront-cash solar installer and have them put solar panels on your roof. Or better yet, why not help your neighbors, too, and set up a community solar program? A community solar project—sometimes referred to as a solar garden—is a solar power plant whose electricity is shared by more than one property.

If you have soil on your land, you can have a direct influence on biodiversity and the carbon cycle. Idle the old fossil lawnmower, cut out the fossil fertilizers and pesticides, invite plants, insects and animals back to your yard, and build your soil into a carbon-sequestering permaculture garden that puts food on your table, too. If that is too much of a hassle, it can also be done as a small community garden or club where together you develop edible gardens, for tastier, healthier, less expensive local fare. The Y on Earth Community empowers people with practical, hands-on information and inspiration to enhance health and well-being while deeply aligning our lives and communities with global strategies for regeneration and stewardship. They have developed videos and community toolkits to help people getting going anywhere.

Some of us hold stocks, bonds, pension plans, or other private savings that may be inherited or self-generated. However big or small your holdings, the same principles apply: All your funds can and should be allocated to so-called impact or responsible investments, terms that entail both social and green placements. Find out what share of your funds are supporting old, dirty growth and disinvest. By reinvesting with healthy growth in mind, we can fund and influence the structure of the economy.

The way you invest matters. The simplest (and least effective) option is to directly buy shares in public traded companies with a sustainable business model, like Unilever, Interface, Tesla, Beyond Meat, Scatec Solar, or Sungevity. If you divest and reinvest by buying shares already on the stock exchange, your money isn't available for the company management to use for investments and growth in impact. So buy and hold has very little influence on the company. If you participate in a new initial public offering (IPO), bond or share issue, however, the money you put in goes directly to management for realizing their business potential. Either way, owning shares also gives investors the chance to raise environmental, social, and corporate governance issues in any company. By filing advisory shareholder resolutions, active shareholders can raise issues at the shareholders' general meetings to the attention of company management, often winning media attention and educating the public as well.

Professional offers for impact investment from main financial institutions are usually reserved for high-net-worth individuals. Small investors had almost no opportunity to get involved—until crowdfunding sites took hold and the number of green, community-oriented, low-threshold investment funds began to grow over the last decade.[14] These platforms allow you to use your money to bring ever more people into the ownership fold. Using Kiva, for instance, you can make a microloan to a woman in Kyrgyzstan so she can buy dairy cows for her organic milk operation—or to a refugee in Uganda so she can buy shoes and sell them to support her family. Kiva microloans bring zero interest to you as a lender but have a very high repayment rate, over 97 percent.[15] You might also find investment funds in your community that allow you to lend money to local businesses with a strong social and environmental ethos. For larger investments, you can seek out companies like Trine, mentioned in chapter 2, and help communities purchase solar energy, aiming for repayment with interest as they save money from buying less kerosene. Trine also pools and then matches peer lending with larger institutional investors, leveraging the sums to substantial amounts that really make a difference.

If you would rather hold a small stake in a new, healthy-growth startup company, there are a number of alternatives, thanks to relatively new crowdfunding legislation. In the United States, Wefunder or Crowdfunder offer investment opportunities in enterprises with social and/or environmental goals. Such equity crowdfunding initiatives are mushrooming all over the world. You do need a certain appetite for risk and losses to enter this field. But if you want your money to make an impact anyway, it's an exciting space. As with every place where there is opportunity for collecting cash quickly, however, frauds and conmen show up and move in. Thus, it takes time and careful scrutiny to avoid the rotten apples.

Finally, of course, if you own assets you should pay your taxes along the way. By cleverly avoiding estate and inheritance taxes, inherited income in particular can be ensconced in funds for generations. This is an important driver of inequality in countries like the United States.

AS CITIZENS: LET YOUR VOICE BE HEARD WITH OTHERS

Consumer, worker, owner—all these roles are important. But it is as *citizens* that we can find our most powerful role in standing up and speaking out to our electives. Most of us live in a democracy, or demo*crazy* as some friends of mine like to say. The truth is, the voting cycles of democracies don't work very well for public long-term decisions, especially when one needs to take bold, quick actions whose paybacks arrive far out in the future.[16] Democracies suffer from the short-term goals of politicians trying to get reelected. They are easily swayed by undue corporate influence. They can be undermined by voters who feel too helpless to show up at the polls. Or taken off-track by those who are ill informed but show up at the polls every time and vote against their interests.[17] In a democracy, change has to be debated and approved at many levels, which slows down action in a crisis. So, someone, which means you and me and we, has to *make democracy work better*. All the time.

Social norms about what it means to be a "good" citizen have shifted over the last decades. Voting, paying taxes, and serving on a jury if asked used to be the core characteristics. Sociologist Russell J. Dalton calls this *duty-based* citizenship, shaped by the zeitgeist in the 1900s. In the 2000s, however, new norms are developing, leading more of us toward an *engaged* citizenship in which we take it upon ourselves to volunteer for organizations, become watchdogs of politics and corporations, and protest in the streets when necessary. We also connect on social media, which has revolutionized the impact of even small groups and allowed social movements to spread rapidly, disrupting old-style power balances. While duty-based citizenship may have weakened since the 1970s and 1980s, *engaged* citizenship seems to be expanding, especially among the young, according to Dalton's research.[18]

This engaged citizen is central to the shift toward healing growth. Without citizens pressuring government and business to transform, democracy will lack both its legitimacy and its guiding force. You can either give up on "democrazy" and go home to the couch, or you can organize, organize, organize. Citizen's Climate Lobby—to just name one out of the million worldwide initiatives and NGOs registered—is fighting hard to build political support across party lines to put a price on carbon, specifically a revenue-neutral carbon fee and dividend at the national level. Putting a proper price on carbon, and then giving every citizen money back as a dividend, is a fantastic climate solution that addresses both footprint and inequality at the same time, in addition to creating millions of jobs by making all the other energy innovations even more attractive.[19]

As citizens, we are of course also townspeople. We're not just individuals inhabiting discrete and disconnected roles like consumer, employee, owner, or voter. Our whole community, large or small, is embedded in the very word: *Citizen* comes from Latin *civitas,* meaning city. Thus, in the citizen, we find the city. In the city, we find citizens. We are in it, and it is in us. This challenges the old Western idea of the individual as a single, self-contained, independent unit of society. To reflect this reality of interdependence, cultural psychologist James Hillman has proposed redefining the individual Self as the interiorization of community, the taking-inside ourselves what is perceived as being around.[20] You will then no longer fall into us versus them thinking but instead feel connected and part of these Others, the fellow citizens, as parts of your Self. Perhaps that is the kind of mindset that will demand and enact the kind of change we need. As Paul Hawken writes: "Healing the wounds of the earth and its people does not require saintliness or a political party, only gumption and persistence. It is not a liberal or conservative activity; it is a sacred act. It is a massive enterprise undertaken by ordinary citizens everywhere, not by self-appointed governments or oligarchies."[21]

It's a complaint commonly heard: There's nothing I can do! The next time that voice rises, whether it's from you or someone else, one good remedy is to think through these four roles: consumer, employee, owner, citizen. Within each there are abundant opportunities. Why not start right now? Put down the book/screen, draw four columns, and fill in at least four (small or big and hairy) opportunities you have under each.

When enough of us start realizing those opportunities, "democrazy" starts inching toward sanity.

WHY WAITING FOR THE RIGHT PRICES WON'T DO IT

But why are innovative yet risky actions by leading companies and extraordinary efforts by engaged citizens needed to achieve system change at all? In the rational economists' worldview, the solution to pollution and resource overuse is quite obvious and easy (on paper): Government puts a proper price on these so-called externalities. With the correct prices on them, pollution and overuse go away as their costs increase. The market will come up with more effective, smarter solutions because that is now profitable, and the sustainable products will become the cheapest. In this economic worldview, the taxes on externalities ought to come into effect immediately after an economic report has thoroughly calculated the correct prices and a commission recommended them.[22] Such taxes and the institutions to perform ecosystem oversight are assumed to simply appear in time—because they do so in the economic model. As soon as the right incentives are in place, corporate and consumer action shifts in the markets, as if by an invisible hand. All the politicians need to do to accomplish this is to have faith in the advice from the economists and legislate accordingly. And hence there is no reason for activists to chastise the companies and paternalize the consumers as to what they "ought" to be doing.

But what this conventional economic view assumes away as *external*, and thus outside to the "circular flow" in the markets (figure III.a), does not happen by magic. In messy, real-world social institutions—like norms, laws, civil sector organizations, affordable housing projects, nature reserves, etc.—can only be built and improved by outspoken and engaged citizen support. And thoughtful effective taxation is only executed—at best—after extended, long-winded multilateral conversations, which may happen way too slowly for sensitive ecosystem boundaries, and for the left-behind children.

Therefore, the economist's ideal solution—proper pricing of the externalities—can be an effective *outcome* of the solution. But simply calculating them correctly with the functions based on efficient markets in equilibrium is *not* the solution itself. Those calculations are at best a first beginning of the solution. So in addition to forward-leaning healthy growth companies and engaged citizens, the third part of the full solution is the long *process* of governance, which finally can lead to responsive regulation and prices.

There are three main shortcomings from the externality-pricing-in-perfect-markets-mindset of conventional economics when it comes to understanding the much-needed role of governments in systems change:

1. The need to shape the supply-side proactively by encouraging early-stage entrepreneurial solutions until they are ready for scale up and can benefit from the economies of scale.

2. The need for public funds to stimulate demand for the smart resource-productive solutions in an equitable way throughout the whole market (not just taxing the bads).

3. The need to build a complex, multilevel, decentralized set of institutions to govern the commons, perform oversight effectively, and adjust regulations and taxes responsively.

That's why the next crucial step is to address how the public thinks and feels about the governance process. How can governments regain the trust of the public so as to work effectively for a system change toward healthy growth?

10 REBUILDING GOVERNMENT: AN ENTREPRENEURIAL, STIMULATING GUARDIAN?

Miles and miles of asphalt on each side of the hotel. Huge ten-lane bridges. Cars hurling past me at seventy miles an hour. As I walked down to the harbor along the mouth of the Los Angeles River at Long Beach, it looked like even the shoreline had been bulldozed and straightened out. The early-evening sun poked in and out of orange and pink clouds far west, partly illuminating the dark silhouettes of harbor cranes.

Then, like Alice tumbling down into a wholly different world, I came upon a little green wetland sanctuary: "Golden Shore Marine Reserve" read a sign hung on a tall, forbidding chain-link fence. "No fishing. No rock throwing," read another. A haunting, flutelike bird call rose above the din of traffic. How weird to hear a sound evoking the abundant wildlife, now long gone, that lived for millennia along this river. A closer read of the first sign revealed the area had been set aside to reestablish some of the natural wetlands that once occupied vast areas around the mouth of the Los Angeles River. "The reserve is fragile," it warned. "Please do not enter."

Fragile indeed. It was surrounded by huge boats, huge buildings, huge cranes, huge bridges, racing cars, everything that seems to make America "great." It seemed incredible that wild birds would find any solace at all in this little green spot. But there stood a heron out in the mud, serenely picking for food as if everything around it was just fine. The reserve is probably not more than a few hundred feet on each side. Someone had pulled washed-up plastic and other trash out of the sand and the grass, leaving it in a pile in the uppermost corner. Walking a bit farther along the imposing fence, I came across the next sign and began to wonder if the plaques might actually outnumber the birds. This little parcel of wetland had been reclaimed from a parking lot, it said, to support spawning fish, absorb floods, and support native plants, insects, and birds.

A pelican flew in low over the steely gray river waves, made a perfect landing on long plastic buoy line, drew its long beak in, and watched me carefully. Could I be trusted? Or should she fly farther out? After a short while she took off as if to say, "Never trust a human."

However much this vulnerable little wetland epitomizes our human toll, I feel grateful for it. It is here, at least—a tiny but clear witness to the dedication to bring back some of the vast wildlife that has been decimated over the previous century or two. Without these patches, we'd be much worse off. Just a few acres of shallow, marshy water can provide a lifeline to birds and sustain small fish and amphibians on the brink. Whoever had the idea to push back the vast parking spaces and secure this patch for our feathery and finned fellows has preserved something key. This whoever turned out to be . . . *government*. A little local branch of it called Long Beach Parks, Recreation and Marine.

In the twenty-first century we'll need to build upon and widen such wise efforts. To bring natural capital back and heal not just our economy but also the growth of wetlands, forests, humus, and kelp. To secure clean water, clean air, and plastic-free oceans. And core, alongside it all, restore dignity to humanity, too.

Walking back to my hotel, I spotted one bowed head beneath the massive bridge. I didn't know who felt the most forsaken: the homeless man sitting under a concrete mastodon ignored by the thousands of cars passing, or that lone, hooting heron making a song no other herons could hear above the steady thunder of passing cars. In economic terms, both natural and social capital had been betrayed by man-made and financial capital. And by a government painfully slow to expand its narrow economic logic.

This chapter does not wade through lists of policy prescriptions. Nor does it review the vast discourse on the relation of government to the governed. Rather, it attempts to reimagine the role of government in relation to markets, capitals, and individual liberty by examining examples of what healthy governance already looks like and focusing on the mental images, or frames, that we—the citizens—hold of government. These cognitive frames have tacit, even subliminal, influence on the kind of policies we support or oppose.[1] When repeated enough to become deeply embedded in the language and minds of millions, they shift democratic participation and elections and influence what kind of governance and public organizations we agree to build. And they can inhibit or enable the public's support for government action to accelerate healthy growth. To regain trust in government, so government can perform its necessary functions, we need to change the language, images, and expectations commonly held about it.

As we've seen, many businesses are already turning to healthier models of growth.[2] Millions of concerned individuals are changing.[3] But unless government, too, shifts from industrialist prerogatives and centralized, rigid, red-tape regulations to smart green stimulus, the swerve of society as a whole may prove to be too little, too late. Rather, the momentum we need comes from societal learning, moving the triangular system toward a connected, virtuous circle.[4] Innovative *companies* can offer products that people want (like "freedom energy" in the form of rooftop solar and batteries, electric cars and buses, or nontoxic local foods from farms that improve the soil). Seeing that attractive, cleaner, and safer products are possible and available, *citizens* will then want more and urge their governments to apply both stick and carrot so that companies will deliver them. When *governments* further improve the frameworks for greener, fairer products, the profit margins from these products pick up and fuel further investments. Leading *companies* then need to collaborate and join citizens' calls for more stringent regulations on the corporate laggards and overcoming their biased lobbying. This closes the gap and gets the triangular system working in a self-reinforcing manner. With persistent pressure from consumers on companies, from citizens on government, from government on companies, and from companies back on government, it can—hopefully—reshape capitalism to work better.

If there is no or little trust in government, though, its crucial role as coordinator and shaper of the triangular market system may become near impossible. Even simple, low-cost policies, such as better insulated windows and buildings, might not take off.

GETTING PAST GOVERNMENT DISTRUST

Those who follow US politics are quite familiar with depictions of Washington as "big government." Bloated, corrupt, bureaucratic, and inefficient. A swamp. Similar complaints can be heard in the European Union about the "Brussels bureaucrats." The words evoke an image of an arrogant, oppressive, and authoritarian central government that takes away people's freedom. Frequently used metaphors reinforce that image: Big government *fleeces* you with taxes through the IRS and *stalks* you with surveillance through the NSA.[5]

Donald Trump won the 2016 elections by framing himself as an "outsider" who would "drain the swamp," conjuring up memories of Ronald Reagan's 1981 inaugural zinger: "Government is not the solution to our problem; government is the problem." Both capitalized on a conservative view that the relationship between government and individual freedom is a tug of war. If the government gains ground, private freedom loses.[6] If individual freedom is to grow, government has to withdraw and shrink. Hence the calls for "smaller" government and the demands for "free" markets.[7] Inside this frame, government becomes your enemy. When such metaphors are taken further, the mind may even be led into paranoia: The deep state is out to get us, rigid bureaucrats are out to stymie the market, and the nanny state is really a totalitarian surveillance state in sheep's clothing.

Not all Americans share these frames, of course. The left side of the left–right polarization has equally deep suspicions and mistrust of how big corporations are exploiting people, screwing workers, ripping up nature, and undermining democracy. In its view, a stronger government is needed to rein in the corporate takeover of society, pushing back on the market and regaining control over the financial sector. But the mindset that's been steadily gaining ground in today's largely conservative world wants government out of its citizen's lives as much as possible and out of their wallets and their businesses even more. That kind of thinking makes it logical to want to cut or evade taxes and force cuts in public spending.

Some conservatives have called this tactic "starving the beast." Taxes are what nourishes government activity. Take away that source of nourishment and government must inevitably shrink, which would be inherently good if you believe government is fundamentally bad. For antitax advocates like Grover Norquist, founder of Americans for Tax Reform, an organization that opposes all tax increases, this is the explicit and ultimate purpose of tax cuts: "The goal is reducing the size and scope of government by draining its lifeblood."[8]

This government-as-enemy frame has been built and spread systematically for decades.[9] In the 1970s, economist Milton Friedman wrote, "How can we ever cut government down to size? I believe there is only one way: the way parents control spendthrift children, cutting their allowance. For the government, that means cutting taxes. Resulting deficits will be an effective—I would go as far as to say, the only effective—restraint on the spending propensities of the executive branch and the legislature."[10] Inside this rhetorical frame, private sector and conservative economists are the responsible grown-ups. Government is the irresponsible spendthrift. Public spending is childish behavior that needs the firm hand of fatherly restraint.

The tides of public opinion have been turning. Pew surveys show that in the early 1960s, trust in government was at almost 80 percent; see figure 10.1.[11] Today, they are at historic lows. In 2019, only 17 percent of Americans said they could trust the government in Washington to do the right thing "just about always" (3 percent) or "most of the time" (14 percent).[12] So whether Democrat or Republican, most Americans are very dissatisfied. Opinion polls from Gallup find the same sinking pattern: In 2017, when asked if big government, big business, or big labor posed the biggest threat to the nation, two in three Americans (67 percent) chose big government. Among conservatives, the big-government response was even stronger: a whopping 81 percent in 2017 and 92 percent in 2013.[13]

But between the 1960s and today, government has generally contributed to better infrastructure, science, innovation, education, and quality of

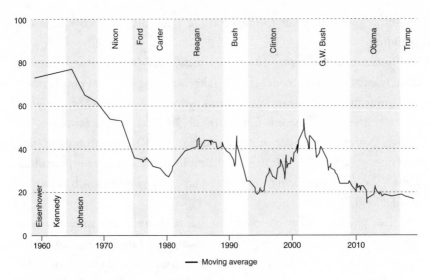

Figure 10.1

Public trust in government is at historic lows. Only 17 percent of Americans today say they can trust the government in Washington to do what is right "just about always" (3 percent) or "most of the time" (14 percent).

life. With the internet and social media, the democratic process may have become more inclusive. So why the distrust? Countries such as Switzerland, Sweden, Norway, and Denmark come out much higher, and rising over time, on confidence in government than the United States, according to OECD statistics and Gallup world polls. Part of the explanation may be that such countries simply are smaller and homogenous population. But that doesn't explain the sinking trend so pronounced since the 1960s, a period of time during which other countries saw rising trust in government; see figure 10.2.[14]

The answer might lie in the unique way in which the image of big government as the enemy of ordinary folks has been cultivated and continually reinforced in the United States. The notion that government infringes on individual freedom, stealing your money to squander it, has been a cornerstone of well-funded campaigns by conservative think tanks and bolstered by thousands of media talk shows and online echo chambers. It has become a core belief for the right.[15] Since Reagan, Republican strategy

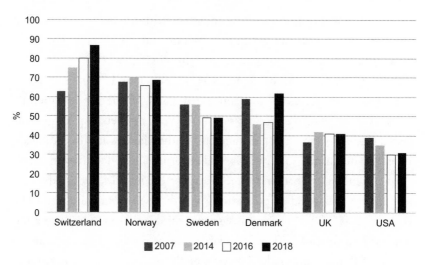

Figure 10.2

Confidence in national government 2007–2018. Data refer to the percentage who answered yes to the question "Do you have confidence in national government?"

involves running *for* government by running *against* government.[16] The resulting scenario has generally gone like this: First, cut taxes on the few rich, but not as much for the many poor. Cut spending on all social programs and federal branches. Then, after starving the beast and doing everything possible to make more public agencies, healthcare systems, and universities dysfunctional from lack of funds—and even risking the shutdown of government altogether—vent anger at federal government for "failing" to work. Proclaim the public system is broken and the private system (that supposedly separate economy) needs to step in and fix it. The reinforcing loop gets even worse. Antigovernment forces profit off the mayhem, taking unfair advantage of loosened regulations, weaker education, or replacing public services with private ones. And then they harvest the resulting voter frustration by blaming and chastising "big government" failures.

Underlying this strategy lies a banal but effective set of mental equations: Government (except for military and police) = bad. Private enterprise (except for those that get subsidies) = good. Private profits = good profits. Government = costs. Corporate executives = effective. Public leaders = ineffective. Receiving money from government = bad profits and bad money.[17] Just countering these equations with the polar opposite (big industry = greed and bad, government = care and good), as is sometimes done on the left, is of course equally oversimplifying.

Rather, work could be done to emphasize that both markets and government are necessary and integrated parts of the same economic *system*. As any reflective economist will tell you, markets require governmental regulations to work fairly and efficiently. Markets are always already embedded in culture, society, laws, and government. It doesn't work when legs declare the arms bad and unnecessary, or the muscles declare the skin their enemy. This kind of polarization is a disease of our minds and our language. It is wildly partial, antisystemic thinking. And it leads to a big problem in a time of crisis when swift action is needed: If more than 80 percent of Republicans and 67 percent of all Americans view government as an opponent in a tug-of-war, then they may not support anything that the government tries to do anyway.

It might seem that government as such hasn't had a good communications department. It has allowed its "brand" to sink and distrust in its services to build in the public mind. Before the government can act transformatively, though, it needs to regain trust—something that involves all of us understanding that it truly matters how we imagine and speak of it. As economist Mariana Mazzucato says: "Imagine how much easier President Obama's fight for US national healthcare policy would have been if the US population knew the important role that the US government had in funding the most radical new drugs in the industry. This is not 'propaganda'—it's raising awareness about the history of technology. In health, the State has not 'meddled' but created and innovated."[18] Cognitive linguist George Lakoff puts it this way: "We, as citizens working though our government, have provided . . . the blessings of modern America. . . . *The private depends on the public*. Public resources make private life possible."[19]

For well-functioning governance in the twenty-first century, then, new images and new metaphors are needed. Just imagine what might happen if the broad majority viewed government as being more *entrepreneurial* than bureaucratic, more *stimulating* than stifling, and more *guardian* of your personal freedom and safety rather than overreaching invader. The

2020 coronavirus crisis irrefutably highlighted all these key characteristics of government.

THE ENTREPRENEURIAL STATE: SHAPING FUTURE SUPPLY

When government and the private sector are portrayed as polar opposites and starve-the-beast frames prevail, then free enterprise, innovation, and individual liberty in markets are actually all weakened. Erase those frames and discussions can shift away from whether or how to shrink government and toward how to enable governments to fulfill their market-designing roles better.

Markets are wonderful, complex, freedom-enhancing systems. But markets don't know when to stop taking natural resources as long as there is more demand. So, without oversight of market transactions, the last fish may be fished, the last tree cut down, and all easily accessible mineral deposits emptied. Markets don't automatically respect the commons unless the relevant players and stakeholders get together and employ tools like trusts, reserves, licenses, or negotiations. And well-functioning tools don't operate themselves. Only an uncorrupted government can set fair caps, frameworks, or impartial rules and enforce them justly. Markets, then, need firm, consistent, trustworthy, and responsive governance around them be vital, creative and free. That's the (apparent) paradox.

Markets must also be wisely shaped. A case in point is the iPhone. Steve Jobs is well-known for starting Apple in a garage and turning it into the largest company in the world by market cap. Its iPhone, with a touch screen, mobile internet, and smart voice recognition, seems to be the ultimate expression of Silicon Valley–style private entrepreneurship. Right?

Well, on closer examination, more or less all of those technologies exist not because Steve Jobs or any other private entrepreneur developed them but because his government did. It paid for research and development until the technologies were ripe enough for private enterprise to run the last leg of the stretch. In fact, the US government has been a leading player in the original research funding not only of the internet but all the other technologies—GPS, multitouch screen display, DRAM cache memory, LCD displays, lithium ion batteries, and the voice-activated personal assistants—that make smartphones a miracle of American technology. It's not the government that did it all. Or Apple that did it all. Government has R&D funds and staying power. Apple has R&D funds and amazing designs. When governments and business cooperate and challenge each other over time, their interfaces connect in unpredictable ways. They stimulate innovation and new products in a multifaceted creative process.

Steve Jobs was no doubt a genius when it came to vision, design, marketing, and commercialization, as were his teams over the years. But his and their genius were given opportunities to blossom and grow by hordes of mind-boggling experiments funded by government sources that none or very few venture capitalists would ever have dared put their money into at the start. Apple successfully put all those components from the technosphere together in an outstanding design and software ecosystem to make world-changing impact.[20]

It's not just the iPhone. Despite the myths about the shale gas boom being driven by wildcatting entrepreneurs operating independently from the state, the US federal government invested heavily in the technologies that unleashed it.[21] Mazzucato points out that in technologies from shale gas to the internet and biotech, "the US State has been the key driver of innovation-led growth—willing to invest in the most uncertain phase of the innovation cycle and let business hop on for the easier ride down the way. If the rest of the world wants to emulate the US model they should do as the United States actually did, not as it says it did: more State not less."[22]

In emerging economies such as China, the public sector is investing heavily in new green technologies, committing huge sums to renewable power sources like solar and wind. The government is also spending heavily on becoming a world leader in artificial intelligence. The government does not view this as a cost. Rather, China sees it as an investment in its engines of future growth, something that will strengthen innovation and export potential and nourish otherwise vulnerable and risky ventures to grow into competitive new businesses. This role of the entrepreneurial state needs to be understood by the rest of the world, and especially the United States, where the dominant political narrative is endangering funding for future innovation and economic growth.

So, it really matters how we envision and speak about the state in relation to markets. Because the metaphors we collectively use will eventually spill over into attitudes and expectations, and show up in polls. Polls then feed the rhetoric in politics, and politicians keep reinforcing these metaphorical framings, strengthening them in our minds. If you hear a thousand times that the state is bloated, the Environmental Protection Agency is overreaching, markets are strangled, taxes are draining, and government spending is irresponsible, lavish, and corrupt, these speech patterns along with their synaptic patterns in the brain are fortified. It shapes the expectations and attitudes you hold of the state. And it will deepen inequality.

In entrepreneurial public–private partnerships, we can imagine the government to be like the skin of a mango that nourishes and protects the fruit inside so that it can grow. You can't have the sweet, healthy flesh of the market without its protective encasing. The skin transports the nutrients, allows respiration, and handles the waste. It gives the fruit freedom to grow and safety against parasites and intruders. The skin directs the growth in a certain direction and expands when needed while also providing clear delineations inside which fair play can be achieved. It is more like symbiosis than a tug-of-war.

In a world turning toward healthy growth, the new job of government is to proactively shape and design the growth of emerging markets. In conventional economics, it has been common to see the state only as an entity that should correct market failures after they have occurred. It is a reactive fixer that steps in when a company forms a monopoly, or cheats consumers, or spews dangerous levels of pollution. To some on the political right, even fixing market failures is a sin, because such attempts would lead to yet worse outcomes in the form of government failures.[23]

The task ahead is to reimagine the role of government as a market designer, co-creator, and shaper, so natural and social capitals are strengthened in an integrated way. In mission-driven innovation, the government can decide in what direction the skin can nourish and grow the market flesh.

As Mazzucato and other market observers point out, the direction that "free markets" tend to follow on their own is problematic because they easily get stuck in path dependency. This means that dominant market players just grow bigger and bigger by following each other in the direction from which they started, all continuing down the same path of the least short-term cost, where they find more economies of scale. They become oblivious to the societal and other effects they may be creating as they grow in their conventional direction. Economists trying to solve such problems have spawned a relatively new field called market design, a system-based methodology that works with a market's unique traits to fix it when broken, or uses systemic analysis to create markets that society needs. It connects well with mission-based innovation. Explains Mazzucato, "When the world is confronted with great societal challenges such as climate change, youth unemployment, obesity, aging and inequality . . . *the State must lead*—not by simply fixing market failures but by actively creating and shaping (new) markets, while regulating existing ones. It must direct the economy towards new 'techno-economic paradigms,' in the words of the technology and innovation scholar Carlota Perez. Usually, these directions are not generated spontaneously from market forces; they are largely the result of strategic public-sector decision making."[24]

Smart, proactive regulations can really stimulate the markets toward healthier growth. Many corporations are recognizing that they need better market regulations from governments. Instead of fighting all regulations, companies such as Nike and Starbucks are calling on government to pass aggressive and innovation-driving carbon legislation.[25]

THE STIMULATING STATE: GROWING GREENER AND FAIRER DEMAND

While the entrepreneurial state refers to bringing new innovations, products, or services to the markets, and thus shaping the direction of supply, the *stimulating state* has its eye more on the demand side. How can the government influence the type of market demand that is conducive to turning the economy around toward healthy growth?

There are three main type of stimulation: public procurement, stimulus packages, and unemployment benefits. As we saw in chapter 8, Walmart, IKEA, and other large companies are beginning to clean up their supply chains, and the same should happen in public procurement. The funds that leave government coffers are truly huge and can push change in the marketplace. Taxpayers' money spent by government on goods, services, and infrastructures such as roads, hospitals, or schoolbooks accounts for 13 percent of GDP on average in OECD countries. However, not all of this money is spent by central government. A large share of procurement is carried out at the subcentral level, with local governments responsible for more than half of public procurement expenditure.[26]

There are many corporate providers of market solutions to adapt to the effects of global warming, improve resource productivity, build clean energy and smart and super grids, protect biodiversity, and help the dispossessed. But those who need these solutions often lack funds to pay for them. To stimulate healthy growth, government can provide earmarked loans, co-funding, and directed green stimulus packages.[27] Stimulus for fair growth can also be achieved by transferring extra money into the hands of those who have little, which usually leads to a direct uptick in the demand side of the markets. Consumption rises, stimulating the market. Some call this the use of helicopter money, as if flying around in a helicopter throwing money out of the window, and as an alternative to the central bank's quantitative easing. If extra money is given to the rich by quantitative easing (as it was during the US bank bailouts that were made to prevent the 2008 Great Recession from turning into a depression) or through tax cuts for the rich, they don't usually spend their money, since they don't need to. A lot of windfall from cheap money and tax cuts is then set aside in savings or invested into rising real estate assets. Increasing income for the less well off, however, has a much stronger stimulating effect.[28] The poor tend to spend more or all of their extra income. Distributing money evenly to everyone in a poor segment of a nation seems to have a lot going for it, if you want to create a fairer, more vibrant economy, where fewer grow resentful at government or business because they feel left behind.

In addition to public procurement and stimulus, there is a third way in which the stimulating state can accelerate demand for the swerve to healthy growth. This is by strengthening unemployment benefits. Such benefits and the bargaining power of unions are one of three key characteristics of the so-called Nordic model, where the labor market is closely connected with the government's welfare system.[29] High workforce participation is achieved by securing high enough incentives to work instead of receiving welfare benefits. Yet these benefits are generous enough to give workers economic safety if they lose jobs. The Nordic model also allows a rather high flexibility in hiring and firing compared to many other industrialized nations. This makes it relatively easier for private sector employers to fire employees, but the governments address this by collaborating with unions and companies for training to get new jobs. In this way employees become more mobile and willing to take riskier and future-oriented jobs. With such overall employment security, workers do not take to the streets and strike if there are layoff programs or when companies are closed. The focus from all parties is on new skills, new education, and the process of getting a new job.

Finally, in Nordic countries there are broad agreements and long-term trust between labor unions, trade associations, and government that the collective bargaining and wage settlements do not set wages too high. If so, the export industry would lose competitiveness relative to other countries, and more jobs would be lost in the longer term. Therefore, trade unions often lead the way in promoting efficiency measures through restructuring industries and laying off people, realizing that improving productivity is crucial to ensure future competitiveness, income growth, and safer jobs. By keeping worker median incomes up and unemployment low, domestic purchasing power is kept robust.

THE GUARDIAN STATE: PROTECTING OUR COMMONS

If there are threats to your health, like smog or pandemics, who fights them? Yes, the government. If there are large corporations exploiting their monopoly powers, who can remove that threat to our freedom? Yes, government. And if someone powerful is closing off access to a river or forest commons, or severely degrading them, who can stop that? Yes—our guardian government, through laws, courts, and enforcement.

In addition to being entrepreneurial and stimulating, the government's new role lies in safeguarding your individual freedom, citizen rights, and access to the commons. The government's role in safeguarding freedom and property rights reaches far back in history. Early eighteenth-century British philosopher John Locke highlighted the need for a guarding government to avoid the "state of nature" in which people would live in fear of stronger others harming them and stealing their property. Despite Locke's recognition of the commons, they lost ground—both in a legal and literal sense—to the primacy of private property in England.[30]

This frame of a friendly *guardian state* is crucial to securing not only people's individual freedom and safety but also a fair access to public commons—such as fresh air and waterways, safe public spaces and transport, the information commons such as the libraries, internet, healthcare services, culture, education, and parks, trails, mountains, wetlands, and forests. Without the guardian state to uphold our individual freedom and access, we may end up in serfdom to monopolistic corporations with failing commons, rather than true market freedom.

The guardian state frame can create more citizen support than the conventional economic lingo and frames of "internalizing the externalities" and fixing "market failures" through the "polluter pays" principle.

While manufactured and financial capital is clearly allocated to an owner, whether private or public, many commons often lack such a clear owner. Healthy growth relies on growth in natural and social capitals, where ownership can mostly or only be common. These are pooled resources and public goods like the flow of clean river water, migratory fish stocks, ideas and information, historic sites, or the cultural charm and attraction of a historic city center.

Therefore, the public management of a balanced portfolio of commons—understood as all stocks that jointly belong to a community—is central if economic policy is to raise shared, broad wealth. To avoid overexploitation, monopolization, or degradation, minimum thresholds for stocks must be defined and guarded.[31] This is exactly what the tiny wetland sanctuary I stumbled upon in Los Angeles is about.

One novel approach to strengthening the commons is to give them, like a forest or nature reserve, legal rights in themselves and by setting up a trust or a foundation to guard them. If they are given legal standing, then one can use the judicial system to guard and protect them against damaging overexploitation. One fascinating example of this comes from New Zealand, where the local Māori tribe had long fought for official recognition of their river as a real ancestor and living entity. The Whanganui is the third largest river in New Zealand. Hundreds of tribal representatives wept with joy when finally, in 2017, their river got its own legal standing written into law. The *Guardian* interviewed the lead negotiator for the tribe, who said: "The reason we have taken this approach is because we consider the river an ancestor and always have. We have fought to find an approximation in law so that all others can understand that from our perspective treating the river as a living entity is the correct way to approach it, as an indivisible whole, instead of the traditional model for the last 100 years of treating it from a perspective of ownership and management."[32] Other places, too, from Bolivia to Bangladesh, have recognized legal rights for nature.

Putting all guardian tasks and responsibilities, however, onto one centralized state is as sure a way to fail in governance as taking away all regulations and rules around a market in a mistaken attempt to make it "free." Neither fully centralized control nor a free-for-all brawl in a 100 percent privatized market can or will succeed in overcoming the well-named tragedy of the commons. In Garrett Hardin's famous essay on this "tragedy," he ends up arguing for the privatization of commons to avoid their overexploitation. But Nobel laureate Elinor Ostrom's studies of how certain groups developed effective economic governance of the commons has empirically demonstrated that group ownership and dedicated institutions can handle conflicting interests and access to shared resources. Not all efforts she studied were always successful. But she and fellow researchers have emphasized the multifaceted nature of human–ecosystem interaction and argued against any singular panacea for all social-ecological system problems. Ostrom nevertheless identified eight effective design directives found across a wide range of communities that manage so-called common pool resources. Let's look at them through the lens of that tiny Los Angeles wetland reserve:

1. Define clear group boundaries with effective exclusion of external unentitled parties. (That's me, an unentitled observer, standing outside the reserve fence, looking in at that heron.)
2. Match rules governing use of common goods to local needs and conditions (Los Angeles locals must not be prevented by the tourist industry or overly strict conservation regulations from being able to jog and swim for free on nearby beaches.)
3. Ensure that those affected by the rules can participate in modifying them. (Hear the voices of coastal communities before changing rules.)
4. Develop a system for monitoring members' behavior and maintaining accountability that is carried out by community members themselves. (Yeah—that recycling system for trash on the beach.)
5. Use graduated sanctions for those who violate community rules. (Fines for littering, leaving dog waste on the sand, shooting wildlife, and so on.)
6. Provide accessible, low-cost means for dispute resolution. (Fight corruption. Provide conflict-transformation approaches in local government.)
7. Make sure outside authorities, including central governments, respect the rule-making rights of community members. (Avoid micromanagement from federal levels.)
8. Use subsidiarity, decisions made at the lowest level possible, to build responsibility for governing the commons in nested levels from the immediate local up to the entire, interconnected system. (Rivers, watersheds, coastal zones, and aquifers may reach over many municipalities, counties, and even states—and hence need institutions with appropriate reach.)[33]

What is needed for a friendly guardian government is to strengthen the quality of social institutions, from courts to tax offices, to encourage fair competition and transparency, and improve compliance oversight.

Improving such governance will also yield a sizable growth dividend from improved social capital. Effective governance can add almost a full extra 1 percent to GDP per person per year, according to recent research from the European Bank for Reconstruction and Development.[34] In this way the commons can be nurtured at all the nested levels of governance, from small communities to the state, national, and international levels. The greatest commons of all are, of course, the air and the climate itself.

WHY IT MATTERS HOW WE FRAME THE GOVERNMENT

In summary, new images, metaphors, language, and practices are required to escape the public-versus-private tug-of-war. Broad citizen support is needed to transform governments from gray growth administrators into healthy growth champions in the twenty-first century. Governments could rebuild trust by boldly taking on entrepreneurial, stimulating, and guardian roles in a people-centered manner. To get the triangular system of free corporations, free individuals, and responsive government going, well-functioning governance is utterly necessary. Competitive markets are great and necessary but not sufficient. Maybe US citizens are right, after all, in naming government as the most important problem facing the country—not because governments aren't needed but because they are entrenched and difficult to change, at least without a wide and broad swell of engaged citizens. The deep reframing of government is in reality a huge job that will take decades for engaged citizens, think tanks and civil society to accomplish. That is similar to the decades that it took neoliberalism to undermine government in people's minds by portraying it as inefficient, irresponsible, and bloated.

A healthy economy consists of a broad set of productive, natural, and social capitals. Private property rights can take good care of the produced and financial capitals. But without governments thoughtfully cultivating a whole suite of natural and social commons at different levels, these commons are likely to continue to wither.

Both the financial and coronavirus crises have demonstrated the crucial need of entrepreneurial, stimulating and guardian government.

But if governments do not reclaim trust by getting institutions working in an impartial way that citizens experience as beneficial,[35] the transition to genuine healthy growth may not happen in time to meet the world's goals for sufficient change by 2050. The overall rates of change may be too fragmented, uncoordinated, and slow. It's therefore time to look ahead, in the final chapter, and see if individual, corporate, and government contributions really can and will add up to the kind of system change we need.

11 DOES IT ALL ADD UP? FOUR SCENARIOS TO 2050

We can, in hindsight possibly characterize 2006–2016 as the years of awakening and conviction by the global business community. Business consultants from McKinsey reported that they "could see how industrial attitude evolved by the month" among their clients. All the input from the dawning renewables boom, CO_2 pricing, the science community, climate extremes, and the Paris Agreement, they noted, "created a broadly held conviction that clean energy systems and resource productivity were major business themes for the coming decades."[1] By 2013, 93 percent of executives said that sustainability was critical to the future success of their companies. In 2007, none of top five global threats were sustainability related in the World Economic Forum's survey. But since 2016 three of five have been—and in 2020 all top five risks were—sustainability concerns.

In the *2020 State of Green Business Report*, Joel Makower, cofounder of the GreenBiz Group, observed, "Companies around the world are moving more quickly than ever to reduce the business risk that comes with these threats to natural capital and human well-being. Indeed, many are moving far faster than their political leaders to make the necessary shifts in how they use resources more efficiently and create fewer waste streams . . . The world's problems may be perilous, but they need not be paralyzing."[2] The health company Novartis, for instance, is spearheading a new financial, environmental, and social impact valuation of their impact on society in addition to economic value. In the *Impact Management Project*, more than 2,000 such enterprises, investors, and practitioners have come together to build global consensus on how we talk about and manage both social and environmental risks and positive impacts.[3]

The biggest obstacles are not technological but lie in government inertia and stagnant organizational leadership. Despite the inertia, some (richer) countries began using fewer metals, fertilizers, fossils, and minerals while raising output. Many (richer) countries also cut CO_2 emissions while growing their economies. Carbon productivity growth is on average picking up speed everywhere. And the carbon footprint embedded in global trade has been falling since 2008.[4]

But does all the current progress add up enough to make a real dent at macro and global levels in light of the multiple, intertwined, and escalating crises of climate, inequality, and biodiversity loss?

Let's not mince words. A host of recent reports has painted a picture of planetary wreckage: on wildlife—the 2019 IPBES Global Assessment Report on Biodiversity and Ecosystem Services; on climate—the IPCC Global Warming of 1.5°C report; on lack of action—the 2019 UN Environment Emissions Gap Report; and on inequality—the 2018 World Inequality Report. Science cannot speak any louder or more clearly as we enter the 2020s. Despite some uplifting progress, there is overwhelming documentation that currently *no* country meets the basic needs of a good life for its inhabitants at a level of resource use that would be globally sustainable. If the natural world in coming decades is to resemble what we grown-ups in 2020 are used to since we were born, and is to be enjoyed by people in somewhat equitable and stable societies, there is no way around a sweeping transformation of the material wastefulness in our economies.

Answering that call to large-scale transformation is the defining challenge of our generation. The current rate of change, as we have seen in prior chapters, is far from enough to address the major existential crisis we face.

THE BOTTOM-UP POTENTIAL IS READY NOW—IN ALL SECTORS

Happily, there are many good reasons for any company to turn around to healthier growth. Consider just the four main factors for modern corporate valuation: brand, performance, risk, and talent.[5] First, a *brand* grows strong and profitable by securing customer loyalty, social license to operate and increased sales. Gaining market share by improved reputation is a key objective to all executives. Second, committing to sustainability practices seems to translate into overall higher operative *performance*, leading to lower costs, less waste, better employee productivity, and improved social impact. Third, by having properly integrated green, fair growth, the external *risks* to the company are reduced. Being exposed as toxic, unfair, or irresponsible can hit revenues, stock prices, and license to operate. Rising prices on resources, carbon emissions, or government regulations can increase costs in the future if the margins are sensitive to these expenses. Fourth, *talent recruitment* becomes easier and less costly as the company is seen as a better place to work.

As the business landscape shifts, the fear of missing out rises, and nothing focuses executive teams like the prospect of slipping behind their fiercest competitors. Anyone who has their radar tuned toward the future can now see where each main sector is heading. The specifics of each solution, of individual companies' positioning, or of the exact market timing may still be blurry. But the overall direction is increasingly lucid.

- *Buildings* both new and old will transform into net zero, plus-, or powerhouse standards, comfortably adjusting lighting and temperature to what its inhabitants require. Increasingly, buildings get smarter and generate more energy than they consume.
- *Transport* will become electrified, self-driving, multimodal, with mobility-as-service overtaking the 1900s car-as-product business model. Ships will become electrified as well, with some adopting liquefied natural gas or hybrid systems before phasing out fossil fuels and running on renewable energy, such as biogas, green hydrogen, and ammonia. Airplanes have lots of potential for energy efficiency as carbon-neutral synthetic fuels, the best biofuels, and electric-hydrogen planes grow their market share. Climate-neutral aviation and shipping industries can become a commercial reality by 2050.[6]
- *Food* will become healthier for body and soil, and travel shorter distances thanks to urban, vertical, and organic agricultural models. New

packaging, processing, and distribution models can cut the need to drive and shop at large malls, reducing overall food waste. Increasingly, the whole food life cycle will move toward circular business models, where nutrients are recovered and returned to growers, regenerating soils. And the current high market share of animal meats in our diets will be increasingly reduced by the development of aquaculture, plant-based, and lab-based alternatives.

- *Industry* and manufacture will go through a robotic and 3-D printing revolution where mass production is replaced with smart mass customization of ever more durable, circular products—improving repetitive jobs by augmenting worker capabilities. Metals, concrete, chemicals, and other materials will increasingly come from circular flows powered by renewable energy, and plastics will be made compostable and biodegradable from organic sources.

- *Waste* solutions will continue to wrench value out of previous "garbage," with smart sensors sorting out the valuable fractions in automatic systems. Gas and water leaks will be profitably regained. And ever more products will be designed with easy, embedded end-of-cycle reuse, including extended producer responsibility, which removes waste itself. Many product materials can eventually be upcycled, making more value each time they are reused, for instance, as raw materials for 3-D printing.

These five end-user sectors give us the basis for a good life. They provide our homes, the ways we get around, the food we eat, and the stuff we use.

Three other sectors enable the necessary shifts in the five core sectors above: energy, finance, and digital. In the coming decades, each one will shift toward zero-emissions, net climate-positive and finally restorative practices. The finance sector so critical to the scale up of this shift has seen already significant changes. The once-reliable return-on-investment stocks of the fossil-fuel sectors seem to have moved into a high-risk, low-margin position. Smart investors—urged even by Wall Street giants such as Goldman Sachs and BlackRock[7]—are shifting billions away from high-climate-risk investments into electric vehicles, solar power, batteries, wind, smart sensors, smart grids, healthy and green urban developments, and other high-growth opportunities that promise higher or safer margins. They can invest in hundreds of attractive, innovative, and transformational solutions that are profitable (or plausibly soon-to-be profitable) in all the important sectors of the economy.[8] In the digital arena, the power grid will gradually be integrated with an internet run by clean, renewably powered server parks, operating our gadgets with safer blockchain solutions and allowing us to use them without worrying about the coal and gas being burned to keep them whizzing and glaring. Together, these three form the infrastructure for a good life: the invisible flows of electricity, funds, and information.

Finally, cities and government are also employing new successful models of governance. China is shifting from being a leading carbon emitter to an inspiring leader in resource productivity, outpacing many "developed" countries.

It seems like we *can* have it all: better lives, smarter infrastructure, better cities, more equal societies, and vibrant ecosystems. There's no lack of bottom-up inspiring examples and solutions that potentially can scale big time. But back to our main question: Does it add up? Is it anywhere near realistic to expect we will deploy the hard work, deep systems thinking, and longer-term commitments necessary to emerge from the grand challenge of the planetary emergency with some dignity? It is quite easy to imagine that humanity simply won't make it. From a darker perspective, most humans are too short-sighted, gullible, violent and self-centered. Well-grounded dystopias abound.

In order to give some science-based answers to what that emergence might require, Johan Rockström, Jorgen Randers, and I brought together a team of scientists and modelers to make the world's first integrated model that could calculate the logical and specific outcomes to the question it all boils down to: How can the world succeed in achieving the seventeen Sustainable Development Goals within nine planetary boundaries—and in time to matter?[9]

A SEARCH FOR ANSWERS IN TOP-DOWN MODELING

The model, named Earth3, is a global systems model linking socioeconomic and biophysical processes. In essence, it is a tool to understand what kind of policies can help the world grow in an inclusive direction while staying within Earth's safe operating space. It builds on more than 100,000 historic and new data points from all over the world in existing well-researched databases. The Earth's biophysical systems, including fifteen potential climatic tipping points, are calculated in an Earth3 submodel called ESCIMO, which is a system dynamics model of global warming toward 2100 and beyond.[10] Our resulting analysis was one of several "Reports to the Club of Rome" investigating the future of economic growth since the *Limits to Growth* book.

Not surprisingly, the main finding is that we can't mimic the past or present to fix the future. Achieving the seventeen SDGs by 2030, or even 2050, with conventional policy tools is not possible. Trying to attain the social SDGs through business-as-usual growth measures, even if somewhat revised and upgraded, leads to a backward slide on the environmental SDGs—with increasing resource use and pollution, and eventually increasing inequality as well. We risk pushing our planetary life systems beyond their tipping points.

That doesn't mean, though, that we have to abandon any of the SDGs. We found it was possible to attain most SDGs within most planetary boundaries by 2050 if world leaders explored and adopted five extraordinary turnaround strategies:

- Massively accelerate renewable energy growth
- Accelerate productivity in sustainable food chains
- Apply new development models in poor countries
- Undertake unprecedented inequality reduction
- Invest in education of women, gender equality, health, and family planning

These five turnarounds can be achieved through economic growth, technological advancements, policies that support inclusion and equity, and partnerships to govern planetary boundaries. It will also require demand-side, behavioral changes, especially in rich countries.

To understand how we arrived at these findings, it helps to understand a bit about Earth3. Using socioeconomic data from 1980 to 2015 for all the world's countries, Earth3 calculates the effects of major socioeconomic developments—in economic growth, population, education, health, resource use, and other factors—on the seventeen SDGs. We use the most suitable publicly available databases to establish the historical trends.[11] The model also assesses the scientific status of global environmental pressures on the nine planetary boundaries. It then uses this information to estimate how many of the seventeen SDGs can be achieved by adopting certain policies in each of seven world regions. And it shows the extent of global pressures on the nine planetary boundaries for different world-development trajectories to 2030 and 2050.

The Earth3 model includes parameters that reflect policy levers in many areas, and that can be seen as a "policy dashboard" for running the world model to 2050. There are levers in each region to influence economic growth rates; jobs, poverty, and inequality levels; energy use and composition; food and agriculture productivity; and education, health, and gender equality. Based on this input, the model then calculates two types of scores: SDG success scores and a score that reflects the state of the Earth's safety margin—our common position within, or outside of, the planetary boundaries.

Each of the seven world regions receives an SDG success score from 0 to 17 for every year from 2020 to 2050, depending on how many of the goals are met.[12] An SDG score is also aggregated for the whole world, weighted by population. To see if SDG achievements are inside the planetary boundaries, we also calculate how these developments impact the Earth's safety margin over time. The Earth's safety margin sums up the risk level of the planetary boundaries. It is scored from 0 to 9, ranked for each of the planetary boundaries.[13] If all boundaries are in the safe zone, the safety margin is 9. If all planetary boundaries are in the high-risk zone, the safety margin is 0—indicating a high probability of irreversible declines in Earth's life-supporting systems and probable societal collapses.

I believe that most of humanity—across all parts of the political spectrum—would subscribe to the vision of fulfilling all SDGs on a safe planet, whatever the population size. So in more specific, measurable, terms, our goal and grand challenge is to have nine billion people achieving seventeen SDGs with Earth's nine life-supporting systems in a safe state by 2050. Using our model, we explored four possible and plausible pathways to meeting that grand challenge by 2050.

FOUR SCENARIOS TO 2050

Each scenario is based on the same historic facts but is shaped by different policy and investment choices over the coming decade(s). We did not assign probability to the scenarios, which means they are not predictions.

The first scenario models how far the world will get on a business-as-usual course to 2050. We call this one *Same*. The second, *Faster*, simulates how far the world could get with faster economic growth. The third, *Harder*, explores the outcomes of pushing known policies harder toward sustainability. The fourth, *Smarter*, calculates the scale of the five turn-arounds actually needed to get there, or plausibly close to it.

Some may consider the business-as-usual scenario most likely and the fourth, turnaround scenario very unlikely. Others the opposite. We hope such foresight analysis can create understanding and a shared language to speak together about the crossroads we're at.

Scenario 1: Same

This baseline scenario explores a future where the *same* policies and actions are applied at the *same* rates of change into the future. Governments and industry will respond to technology, inequality, and climate disruptions in the conventional ways and with the same tempo as they have for the last three decades. Despite rapid technological changes, in particular digitalization, the data from the last decades show that most rates of socio-economic change are slow. In a more-of the-same world, there is perhaps more *talk* about sustainability and SDGs. And many more United Nations conferences. But in practice nations still continue unperturbed. This proves insufficient to deliver on most SDG targets by either 2030 or 2050.

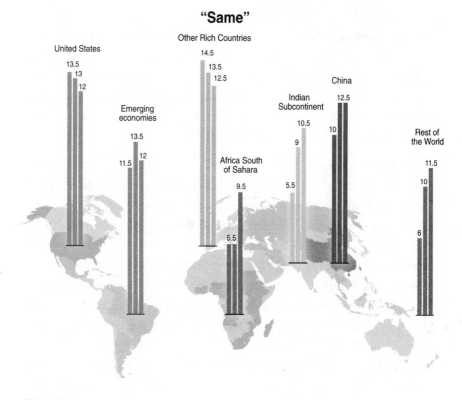

Figure 11.1

SDG success score per region in the Same scenario. For each region, the bar on the left represents the 2010 SDG score, the bar in the middle 2030, and the bar on the right 2050.

The good news is that both hunger and absolute poverty are finally eradicated by 2050. However, the economy's ginormous resource use and waste flows push more planetary boundaries into the red zone. This leaves many of Earth's life-supporting systems in a perilous state with a high risk of irreversible decline. Most people's prospects for further well-being, particularly the poorest 40 percent, will be bleaker in 2050 than today.

In total, the world's SDG score only improves from 8.8 in 2015 to 11.3 in 2050, as figure 11.2 shows. At the same time the Earth's safety margin falls further from 4.5 to 3.[14] The main two reasons are that, first, inequality continues to grow within regions and countries and, second, total human footprint is too high, despite the fact that many countries are dematerializing. Planetary boundaries in high risk zones, along with failing to achieve on SDGs for climate, life on land, and in oceans (SDGs 13 to 15), set all regions on a downward slope from 2040 to 2050. The richest 10 percent will probably congregate in gated communities and pay for their own security forces. By responding to our new problems in the same conventional ways, most people on Earth end up in an even more precarious state in 2050 than we are in 2020.

Scenario 2: Faster

This scenario explores what happens if governments and investors around the world ramp up conventional approaches and succeeds in delivering faster economic growth than expected all the way to 2050. The pace is faster, but the tools are still mainly the conventional, grayish type—such as increasing trade, consumption, digitalization, innovations, and investments and keeping corporate taxes and interest rates low. We ran the model with growth rates that are 1 percent higher in GDP per person per year than the trend used in Same. This makes the global economy significantly larger by 2050, than in Same.

On the plus side, higher incomes become available to solve the world's problems, providing extra funds for more education, clean water, food, more jobs, and elements of the other SDGs for all people as well. But— maybe surprisingly—the high-growth pathway only delivers a little bit better on the weighted SDG success score, from 8.8 in 2015 to 11.7 in 2050 (compared to 11.3 in Same). Simply having higher economic growth of the same kind does little to improve well-being as measured by SDG achievement. Many more people get even wealthier, but societies suffer yet more destabilizing inequality. And the planetary boundaries are pushed yet more deeply into high-risk zones. The expansive economy grows beyond Earth's safe operating space by overexploiting nature's life-supporting systems. Earth's safety margin comes down from 8 out of 9 in 1980 to just 3 in 2050 (see figure 11.2).

Scenario 3: Harder

What happens if the world's decision makers and governments put dedicated effort and dogged determination into achieving the SDGs? Rather

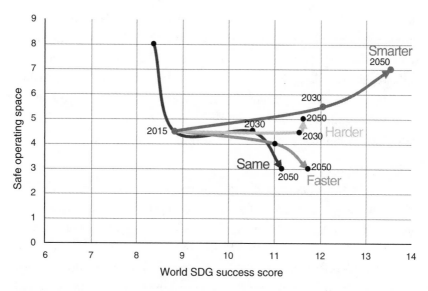

Figure 11.2

Is humanity achieving the SDGs within planetary boundaries? The horizontal axis shows the world's SDG success score (which can go from 0 to 17), and the vertical axis shows the safety margin of Earth's planetary boundaries from 0 (all are in high risk) to 9 (all PBs are in safe operating space). All scenarios show 1980–2050. Only one of the four scenarios rises to the grand challenge of improving the world's SDG success score without eroding Earth's safety margin.

than relying on the invisible hand of faster market-led growth, can stronger determination to deliver on the sustainability goals meet the challenge? In this scenario, governments allocate more funds to pay for more education, clean water, food, jobs, and the other SDGs for all people everywhere. They get their act together, strengthen their conventional policy tools, and raise taxes modestly. Starting in the early 2020s and up to 2050 they get on average 30 percent more rapid progress on SDG achievement per dollar of GDP than they did in the previous thirty years. More workforces, more renewables, more digitalization, larger projects, and more private-public funds are redirected to help achieve SDGs and reduce the pressure on planetary boundaries. These are Herculean governmental and industrial undertakings.

However, by delivering on the SDGs one by one in a piecemeal way, project by project, sector by sector, and ministry by ministry, there are many trade-offs due to the lack of systemic changes. By 2040, the planetary boundaries are still under strong pressure, which leads to stagnant SDGs scores from 2030 to 2050 (also shown in figure 11.2). And many countries still struggle with destabilizing inequality that undermines the sustainability policies. All in all, the Harder pathway leads humanity to undermine Earth's life-supporting systems, if less so than in Same or Faster. Trying harder still does not amount to a worldwide systems change.

Scenario 4: Smarter

The Smarter scenario is a challenge-and-response scenario, in which "creative minorities"—groups of responsive changemakers and activists—in the face of the overwhelming challenges devise solutions that lead to reorienting their entire society.[15] It describes a pathway where people do not just try harder along the well-known tracks but rise to do whatever is necessary to succeed. Rather than copying the same late-1900s type of solutions, growing faster or trying harder, this scenario explores what working smarter entails. It describes the ambitious rollout by most societies and economies around the world of those five transformative measures—bold, connected missions embraced by companies, citizens, and government. In this scenario, world leaders, supported and pushed bottom-up by their engaged citizens and companies, first acknowledge the massive scale of the grand challenge ahead. Then, they throw themselves into the execution of the necessary turnarounds:

1. Massive renewable energy growth must replace enough fossil fuels to halve carbon emissions every decade from 2020.
2. Food chains must be rebuilt from soil to table through less food waste, more plant-based diets, regenerative soil practices, and nutrient cycling. Resource productivity must be improved by an extra 1 percent per year from 2020 to 2050.
3. New development models for rapid economic growth in the poorer countries inspired by characteristics of governance models like China, Ethiopia, or Costa Rica.
4. Active inequality reduction to ensure that the richest 10 percent take no more than 40 percent of national incomes.

5. Investment in education for all, particularly girls and women, gender equality, health, and family planning, which stabilizes the world's population below nine billion.

In the Smarter scenario, these five broad turnarounds are first met with criticism and pushback for being too radical. But starting in the 2020s, in an unprecedented surge where the triangular system of citizens-business-government reinforce each other, the five turnarounds are widely adopted, accelerated, and scaled over the coming decades. The five are all system-wide levers that can deliver on several of the SDGs simultaneously. This leads to high and stable annual improvements in both resource and social productivity. The name Smarter refers to this systemic approach—as opposed to a piecemeal, goal-by-goal approach that often generates trade-offs.

The Energy Turnaround

The first turnaround in the Smarter pathway breaks with the fossil past and turns toward an electric future: While power generation goes renewable, transport, heating, and cooling are electrified worldwide. A massive scale-up of mainly solar and wind power is the enabler on the supply side, while radical energy efficiency brings down end-user demand. Solutions for distributed energy storage, electric vehicles, heat pumps, and necessary distribution infrastructure are all digitized and integrated in smart grids to replace fossil fuels. Nearly all investments in fossil fuels (a historical average of 1.5 to 2 percent of GDP per year)[16] are shifted to renewables, storage, and power infrastructure during the early 2020s.

This historic shift is driven by a combination of pull and push. The market pull happens because renewables can now deliver energy in most domains cheaper than fossils can.[17] "Everyone" wants it, so demand explodes. The push comes as governments introduce tougher regulations with CO_2 pricing, implement renewable portfolio standards (mandating that more energy comes from renewable sources), and cancel fossil-fuel subsidies. This results in several doublings of the annually installed amounts of new wind, solar, other renewables, and storage during the 2020s while demand for fossils starts plummeting.

Many countries also implement bans on new fossil-fuel investments, including announcements during the 2020s of coming bans on sales of new fossil-fuel cars. Most regions adopt some form legislation requiring them to halve carbon emissions every decade, starting with 2020.[18] This aggressive emissions reduction rapidly reduces global carbon emissions and at the same time eliminates human suffering from air pollution by spreading affordable electricity to cities, slums, and remote areas. China takes the global lead, with strong policies for transforming coal reliance to low-cost distributed renewables and electric mobility that make it more profitable for other countries to follow. The direct use of fossil fuels in buildings for heating and cooling is replaced with, insulation, heat pumps, and smart system redesign.

This energy system transformation effectively weans the world off fossil fuels and delivers on the clean energy goal (SDG 7). There is, finally, "power to the people." Giving nearly all nine billion access to enough clean energy for low- to middle-class lifestyles creates a functioning energy democracy, which improves the development of many other SDGs (1, 2, 6, 8, 9, 11–13): it provides better access to light, education, clean water, and communications. In addition to reducing climate change (SDG 13) it helps fight poverty (SDG 1) and make more jobs (SDG8). It makes innovations and infrastructure (SDG 9) more available and reduces food waste and hunger (SDG 2) by access to refrigeration and logistics. It helps make city air cleaner (SDG 11) by replacing combustion. In sum, universal access to abundant, cheap, and clean electricity changes everything!

As a consequence, global carbon emissions fall in the Smarter scenario from almost 40 $GtCO_2$ in 2020 to 20 in 2030, 10 in 2040, and just 6 in 2050. The annual capro is greater than 7 percent per year from 2020 to 2050, and the 1.5°C target is still within reach.

The Food Turnaround

The second turnaround happens in food and agriculture. Smarter sensors, logistics, and storage drive down food waste at the demand side, as well as fertilizer and pesticide overuse on the supply side. People shift their diets to more plant-rich foods, which lowers the share of meat per person—particularly in the richer countries.[19] The food system gets more direct links between producers and consumers, improved packaging, and logistics of easily available, affordable, and nutritious foods that people actually need and want from soil to table. This brings down food waste along the previously overly long food chain of highly processed foods.

New food solutions build on the rapid digitalization of agriculture. Cheap sensors, drones, satellite monitoring, and the Internet of Things make real-time big data available to monitor the state of each field, river, crop, and shop. Armies of small electric robots in the fields replace heavy tractors and pesticides. Through better water management, total water use is brought within ecosystem boundaries. Intelligence embedded in water pipelines helps stop water loss from leakages and secures good water management in all river basins. It makes freshwater pricing more accurate and feasible, giving incentives for better water efficiency. We get more human nourishment from less land, reversing the expansion of agriculture into wild areas and forests. Biogas and composting of organic wastes replace landfills, incineration, and surface runoff to the oceans, creating the capacity to recapture nitrogen and phosphorus and circulate these nutrients within bioregions.

These kinds of both low-tech and high-tech solutions enable increasingly regenerative agriculture to produce more food in better soils without any land expansion. The annual release of bioactive nitrogen starts to decline, decreasing algae blooms and eutrophication. With less need for arable land, forests start to grow back. And climate-smart agriculture becomes a net carbon sink and draws down around 5.5 billion tons of greenhouse gases into the soil per year from 2040.[20]

A less wasteful and more productive food system can also increase people's health as they get more nourishing and affordable food (SDG 2). With recycling of nutrients, it also improves clean water (SDG 6), responsible consumption (SDG 12), and reduces the pressure on climate change, life on land, and life below water (SDG 13–15). In sum, all these improvements lower the footprint of the entire food chain by an extra 1 percent per year relative to the Same scenario.

The Turnaround to New Development Models

The third turnaround involves significantly higher growth rates in the world's poorest countries through new growth models. During the 2020s, the gap between the poorer and richer countries begins to close. The poorer countries increase investments (particularly in infrastructure), strengthen social institutions, fight corruption, and allow favorable trade arrangements in the early stages of their industry development. They are inspired by and learn from relevant characteristics of entrepreneurial governments, such as China, South Korea, Ethiopia, and Costa Rica. First Japan, then South Korea, Singapore, and China managed to quadruple the GDP per person over thirty years. China has achieved an unprecedented duration of sustained economic growth and lifted hundreds of millions of people out of poverty in the process. Ethiopia has seen annual growth rates around 10 percent per year during the 2010s. As other poor countries pull off similar feats during the 2020s and 2030s, they start providing their citizens with reasonable standards of living.

The "China model"[21] of decision-making is preferred in the coming decades by many such countries over the conventional "Washington consensus." The latter prescribed policies such as macroeconomic stabilization, less public expenditure, rapid economic opening with respect to trade, finance, and investment, and the immediate expansion of market forces within the domestic economy, before offering support across borders. But during the 2020s, many of the world's poorer countries prefer to follow the China model to find ways to roll out forward-looking protectionist policies, too. Their entrepreneurial government will protect infant industries in early stages by limiting their exposure to trade competition with advanced global corporations in their own home markets.[22] The effects in these countries are more rapid inclusive economic growth that lifts many millions more out of poverty faster, and also delivers on hunger, better health, education, clean water, jobs, and infrastructure (SDGs 2, 3, 4, 6, 8, 9).

The Inequality Turnaround

The fourth turnaround deals with unemployment and inequality. During the early years of the 2020s, a series of political crises are fed by broad discontent and public protest about the extreme unfairness of wealth inequality. Public awareness along with more responsive government starts lifting median wages. And more progressive taxation succeeds at reducing inequalities in disposable incomes. Many developing countries intensify domestic resource mobilization by improving their tax systems, transparency and closing access to overseas tax havens. As a result, more funds are available for better public service delivery and development for the majority.[23] There is also growing public awareness of the ever stronger economic recommendations to reduce inequality.[24]

By 2025, voters increasingly recognize that social stability is best served by ensuring that the 10 percent richest take no more than 40 percent of income. Downward redistribution of wealth and incomes through policies such as higher unemployment benefits, universal basic income, a shorter working year, and more from the toolbox in chapter 6 are seen as the best ways for businesses and banks to guarantee a stable economic future in the developed world. This works because it puts more money into the pockets of the poor. It allows the less well-off to spend more, which also improves conditions for business, investors, and the banking sector.

The funds raised by progressive taxation of income, inheritance, and wealth are increasingly used for inclusive well-being by delivering on SDG achievement (particularly on health, education, infrastructure, sustainable cities, and responsible consumption—SDGs 3, 4, 9, 11, and 12). The historic trend of stagnant median incomes since the 1980s is reversed around 2025. This starts to regain more trust in government and more stability of policies. That gives opportunity to strengthen the growth of social institutions (SDG 16) and partnerships for the goals across national borders (SDG 17).

The Gender Equality Turnaround

The core of the fifth turnaround is educating all women and investing in gender equality. Funds from progressive taxation are earmarked especially for women in developing countries. This gives women broader opportunities for autonomy and work. In addition, better family planning and urbanization give women more freedom to choose what kind of life and how many children they want. The more female leaders the world gets, the more women become empowered to take positions of leadership, a self-reinforcing loop.

While women worldwide were slowly closing the gender gap before 2020 in critical sectors such as health and education, significant gender inequality persists in the workforce and in politics. The rate of progress for women starts slow, too. Between 2006 and 2016, for instance, the proportion of female leaders increased by only 2 percent.[25] But when women are better represented in leadership roles, more women are hired across the board. More countries follow the example of Norwegian law, which requires a minimum of 40 percent women on the boards of publicly listed companies. This trend picks up speed in 2025, when the world recognizes that encouraging more female leadership is one of the main levers for increasing gender equality in the entire workforce as well as accelerating economic growth.[26] Results are indisputable and raise further public awareness. By the 2030s, it is becoming widely acknowledged that a good gender balance is much smarter and more profitable than the conventional male-dominated networks (SDG 5, 8, 16).[27]

For women, the five factors of education, urbanization, job opportunities, family planning, and reproductive health combine for better well-being, including for their children. This results in women choosing freely to have fewer births, slowing population growth. Surprisingly, global population reaches a peak before reaching nine billion and then starts to decline.[28]

Despite Widespread Conflicts, Smarter Practices Prevail

Regardless of positive trends in many areas, environmental stresses like air pollution, droughts, wildfires, storms, floods, and high tides have been worsening for many decades at least since 1980s. Even in this Smarter scenario, in the decades up to 2040, such extreme events cause more urban crises and waves of migration. They contribute to conflicts, and in some places civil wars. Such conflicts put severe pressure on many fragile institutional structures. Political crises, corruption, and distrust of interventionist government cause periodic opposition to the active entrepreneurial and guardian government roles key to rolling out the five turnarounds. The increased progressive taxation to reduce inequality is also a hotly contested topic for decades.

The world that Smarter describes is still messy and far from a utopia. Social systems tend to reproduce themselves. So deep system change don't come easy. Ever.

Yet, since there are better redistribution measures in place that benefit large majority groups, along with a stronger taxation base and an international commitment to peace and partnership that remains resilient and often responds rapidly, the worst crises are dealt with before descending into full collapse of large cities or more failed states. Among large investors and private companies, there is a rapidly growing realization that large businesses cannot succeed in societies that fail. Increasingly, the financial sector and capital markets start using environment, social, and governance criteria (ESG)—essentially a healthy growth compass—to guide investments. From a feeble start around 2015, ever more of the world's funds start to rigorously *apply* the Principles of Responsible Investment (PRI), integrated reporting (IIRC), climate risk disclosure (TCFD) and ESG analysis. As the ideas behind this confusing soup of acronyms become standardized and move into mainstream, the information is increasingly priced into risk and capital markets. Mainstream investors include these considerations in their day-to-day operations. More than half of the world's wealth gets invested in line with revised and effective PRI and ESG guidelines by 2030.

In the Smarter scenario, the improving trends of education access, unemployment benefits, job sharing, fairer wages, and extensive redistribution make societies more equitable. Also, due to digitalization, end-user efficiencies, circular material flows, and responsive regulations, the extraction and resource throughput falls while the economy grows. Societies are learning how to eliminate wastefulness to get a lot more with much less stuff. The pressures on natural resources and ecosystem services are gradually relieved and reversed as regenerative practices spread. They are also discovering that both the social productivity and the resource productivity feeds back from better social and natural capital assets to give a positive impact on GDP per person growth, the first by stimulating demand, and the second by better resource use on the supply side, cutting raw material extraction, transport, and cleanup costs.

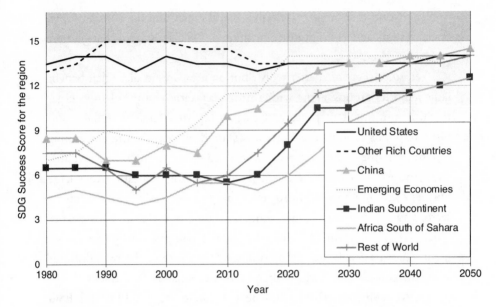

Figure 11.3
SDG success score per region in Smarter from 1980 to 2050. In the Smarter scenario, the poorest regions catch up earlier than in Same, Faster, or Harder. This is the result of the five transformational actions having systemic effects on several SDGs at the same time. The world's total success score thus goes higher. The dark shaded area is the ideal score > 15, the light shade is yellow > 10, below which are unsatisfactory levels of SDG achievement.

By 2050, most regions of the world are delivering on nearly all SDGs (at a global average of 13.6 out of 17). Both India and sub-Saharan Africa have shown tremendous progress (from a 5.5 regional SDG success score in 2010 to 12.5 in 2050, as figure 11.4 shows). This is not a world free of troubles: Many social conflicts remain, and four planetary boundaries are still in the yellow boundary zone (global warming, land conversion, air pollution, chemical pollution). Yet the Smarter pathway seems to point the world's economy in a prosperous direction within the Earth's safe operating space.

SO DOES IT ADD UP?

Will we truly achieve the Sustainable Development Goals within planetary boundaries by 2050? After reviewing these four plausible futures, the

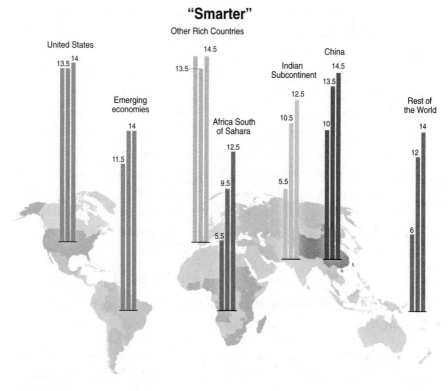

"Smarter"

Figure 11.4
SDG success score per region in the Smarter scenario. For each region, the bar on the left represents the 2010 SDG score, the bar in the middle 2030, and the bar on the right 2050.

honest answer is: we don't know. These scenarios are all based on real-world facts, up to today. The world can continue on its dead-end track of Same or choose to work for and achieve the five turnarounds in Smarter. The deep uncertainty is real, and the future is wide open.

The Same, Faster, and Harder scenarios can all somewhat improve the world's SDG achievement. But they tend to do that at high cost to the stability and risk level of Earth's life-supporting systems. Most of the known tipping points in Earth's climate system are modeled in Earth3. None of them show a definitive catastrophic collapse by 2050. But in the first three scenarios, humanity moves into a very high—and really intolerable—level of risk for having already *ignited* the slow-moving but irreversible declines by 2050. This is illustrated by the downward-pointing endings of the curves in figure 11.2. The natural world gets gradually grayer, burnt-out, flooded, more unstable, and impoverished. More of the same, even if we try harder or work faster at it, simply results in rates of change too slow to register as adequate direction or momentum on the healthy growth compass

Of these four possible scenarios, only the five systemic turnarounds in the Smarter scenario can keep growing the world economy in an inclusive manner while keeping (almost all) planetary boundaries in safe zones. It doesn't get us all the way there by 2050. It shows an SDG score of 13.5 out of a theoretical max of 17, and safety margin of 7 out of 9. This type of transformative, healthy growth can result in an *almost safe* operating space for all of humanity before midcentury. If so, the world's societies can continue to flourish into the future far beyond 2050.

A DEEP REFRAMING: OUTGROWING OUR OLD GROWTH MODEL

The new direction of growth described in Smarter also requires a deep psychological reframing. Growth can no longer be understood mainly from the human-centered perspective of Enlightenment economics, in which all biological life, from soil to sea, is fodder for financial flows. That vision of economics and its accompanying metaphors are deeply rooted in eighteenth-century mindsets. These considered the "lower" domains of vegetative and animal life—along with the cultures and lands of the First Nations—there for the taking (and wasting) by higher, rational minds.

Healthy growth requires embracing a more complex balancing of our long-term productive, natural, and social capitals—as the Smarter scenario shows is possible. As we unlearn the old frame in which growth is a linear increase of annual monetary flows alone, we can foster the new frame of cyclical complexity, an "enlivenment economics." Here financial capital in complex ways starts to also enrich the many networks of natural and social capitals in which good human lives are embedded. This new worldview benefits financial capital, too, over the long term: There is little business and no dividends in a collapsing world.

Healthy growth, if executed in a consistent and integral way, is a credible, profitable solution to our current economic, social, and ecological crises. It can turn *gross* domestic product into *healthy* domestic product, transparently and objectively, year by year. The fight *against* growth can be put to rest.

Our problem is not primarily technological or physical. The solutions we need are available. Earth is still bountiful, bathing in energy from that giant, abundant, free-for-all fusion reactor in the sky. Our problems lie in the way we think and talk about growth. Attack growth as the enemy, and you set yourself up to fail. Because, for a majority, growth subconsciously and archetypally equals betterment. We need to heal our ideas of growth, and that healing starts in our minds.

As a psychologist I don't believe in the puritan urge to arrest the lust for wants and growth, to deny people their inner innate drive toward something "more." Rather, it's time to go along and shape that into a quest for a new course, something more alive, more "healthy." The psychological task now is to reshape the growth urge that has held us in thrall to an old linear, industrial version of it, rather than trying to suppress it. We can, individually and collectively, align ourselves with that deep inner yearning in all living things—from trees to cubs to humans—for growth in cyclical complexity, toward a fuller, more intertwined and inclusive Self-realization. Enlivenment economics view our goal as leading a fuller life, writes philosopher Andreas Weber. "If we adopt this perspective, we will begin to see that something is sustainable if it enables *more life*—for myself, for other human individuals involved, for the ecosystem."[29]

And there's no more time to lose. As the metrics presented in this book show, the "speed" of system change can be measured with precision, for any entity and any period starting yesterday. We know what we need to do: raise resource productivity everywhere by at least 5 percent per year (and preferably more than 7 percent carbon productivity). If any year we do not achieve these rates of change, the efforts can redouble the next, to help against going deeper into the red risk zone. And, for countries with an inequality Palma ratio above 1.5, raise the social productivity of growth at least 5 percent per year in order to move sufficiently toward equitable societies. That means to choose, every year, the healthy growth course on the growth compass: $rp > 5$ percent and $sp > 5$ percent. Real system change lies in that direction. Fully possible. Fully doable. Feels enlivening. Requires determination. And a growing coalition of the willing.

The current state of affairs gives little reason to stay put or satisfied. Succeeding with healthy growth at sufficient rates of change year after year is by no means a given. It even seems unlikely if you look only at the past trends. The first three scenarios that describe ordinary, plausible pathways forward do not get us where our economies need to go.

But I choose to believe we—meaning you and I—can grow in the direction of doing extraordinary things each day, things that break with the stale economic patterns and narrow growth mindsets of the previous decades. Each week brings another opportunity—each month, each year, and each decade—to grow more fully alive. We can leverage the tremendous forces available in mercurial markets, the limitless creativity of people, and the exponential growth of emerging solutions to drive through quicker and more wide-reaching changes than we are able to imagine today.

With this kind of growth, we can still make it to that more beautiful and fair future world.

Acknowledgments

First thanks must go to my colleague and friend over many, many years: Jorgen Randers. Without his sharp analytical take and our fruitful discussions and disagreements over the last decades, I could not have so sharply defined the healthy growth approach.

Joni Praded has—as book editor—supported and refined the project over the last years of writing. She's simply the best editor I've ever worked with, always identifying, cutting, or rewriting those weak spots. Of which there were many. Any remaining shortcomings and errors are fully mine.

A number of wonderful people have contributed thoughts, criticism, and other feedback, among them:

Mads Greaker, John Elkington, Chad Frischmann, Marit Sjøvaag, Espen Skaldehaug, Per Ingvar Haukeland, Daniel Erasmus, Caroline Ditlev-Simonsen, Carlo Aall, Jan Bråten, Sylvia Weddegjerde, Olav Bjerke Soldal, Anne Grethe Henriksen, Erlend Guldbrandsen, Bjørn Vidar Vangelsten, Andreas Fagernæs, Markus Lindholm, Monica Mee, Sandra Sotkajærvi, Katharina Bramslev, Torbjørn Wilhelmsen, Sigrun Aasland, Sigrid H. Melhuus, Helene Lillekvelland, Ida P. Berre, Christine Lundberg, Karin Sandgren, Britt Nilsen, Marie Winsvold, Ingrid Møller, Janicke Garmann, Lars-Erik Aas, Kristin Solberg Willoch, Kiran Kumar, Goncalo F. Farinha, Fredrik W. Mowinckel, Kristin Svendsen, Ketil Stoknes, Martin Julseth, Hallvard Surlien, Cecilie Staude, Ingrid H. Warner, and Jonas Mosskin.

Then there are all my Executive students at the Norwegian Business School over many years. Deep appreciation goes out to all of you—as well as those possibly not mentioned.

Thanks to the MIT Press team, Beth Clevenger in particular, and two anonymous peer reviewers. Thanks to R. Jamil Jonna and *Monthly Review Magazine* for the permission to use figures 4.2a and b. Thanks to Amory Gethin and the *World Inequality Database* team for the permission to use figure 6.1. Thanks to Wilkinson and Pickett and the *Annual Review of Sociology* for the license on figure 6.2. Thanks to Kristoffer M. Hansen at Leidar, and Riccardo Pravettoni for help with the design of other figures.

Finally, the deepest gratitude to my very closest supporter and critic, Anne Solgaard. Without you . . .

Notes

PREFACE

1. Carl Jung, *The Collected Works of C. G. Jung*, vol. 10, part IV, "The Undiscovered Self (Present and Future)" (Princeton: Princeton University Press, 2014).

2. As since confirmed and exposed in Johann Hari, *Lost Connections: Uncovering the Real Causes of Depression—and the Unexpected Solutions* (New York: Bloomsbury, 2018).

3. James Hillman, *We've Had a Hundred Years of Psychotherapy—and the World's Getting Worse* (San Francisco: HarperSanFrancisco, 1992).

INTRODUCTION

1. Max Roser, "Economic Growth," OurWorldInData.org (2019), https://ourworldindata .org/economic-growth and https://ourworldindata.org/extreme-poverty, accessed May 2, 2020.

2. Benjamin M. Friedman, *The Moral Consequences of Economic Growth* (New York: Vintage Books, 2006), 1.

3. Growth proponents along these lines of reasoning include Lant Pritchett and Lawrence H. Summers, "Wealthier Is Healthier," *Journal of Human Resources* (1996): 841–868; Paul A. Samuelson, "Optimum Social Security in a Life-Cycle Growth Model," *International Economic Review* 16, no. 3 (1975): 539; Paul A. Samuelson and William D. Nordhaus, *Economics*, 19th ed. (Boston: McGraw-Hill Irwin, 2010); Lawrence H. Summers, "Age of Secular Stagnation: What It Is and What to Do about It," *Foreign Affairs* 95 (2016): 2; World Bank and Commission on Growth and Development, *The Growth Report: Strategies for Sustained Growth and Inclusive Development* (Washington, DC: World Bank on behalf of the Commission on Growth and Development, 2008).

4. "Towards the End of Poverty," *Economist* (June 1, 2013), https://www.economist.com/ leaders/2013/06/01/towards-the-end-of-poverty; Jeremy Warner, "Is Economic Growth a Friend or Foe to the Environment?," *Telegraph* (September 26, 2019).

5. Facundo Alvaredo et al., *World Inequality Report 2018* (Cambridge, MA: Belknap Press, 2018).

6. For a recent review of the so-called Easterlin paradox, see R. A. Easterlin et al., "The Happiness-Income Paradox Revisited," *Proceedings of the National Academy of Sciences* 107, no. 52 (2010): 22463–22468; Richard A. Easterlin, "Happiness, Growth, and Public Policy," *Economic Inquiry* 51, no. 1 (2013): 1–15; Richard A. Easterlin, "Paradox Lost?," *SSRN Electronic Journal* 16, no. 2 (2016).

7. David Suzuki quoted in William Rees, "David Suzuki Is Right: Neoliberal Economics Are 'Pretend Science,'" MAHB, May 10, 2018, https://mahb.stanford.edu/blog/neoliberal -economics-pretend-science/; George Monbiot, "Growth: The Destructive God that Can Never Be Appeased," *Guardian*, November 18, 2014, https://www.theguardian.com/ commentisfree/2014/nov/18/growth-destructive-economic-expansion-financial-crisis; Smil is quoted in Jonathan Watts, "Interview: Vaclav Smil," *Guardian*, September 21, 2019, https://www.theguardian.com/books/2019/sep/21/vaclav-smil-interview-growth-must -end-economists; "Transcript: Greta Thunberg's Speech at the U.N. Climate Action Sum- mit," *NPR*, September 23, 2019, https://www.npr.org/2019/09/23/763452863/transcript -greta-thunbergs-speech-at-the-u-n-climate-action-summit; Jason Hickel is quoted in "Our Addiction to Economic Growth Is Killing Us," *Ecologise.in*, August 14, 2017, https:// www.ecologise.in/2017/08/14/jason-hickel-bbc-addiction-economic-growth-killing-us/; Naomi Klein, *This Changes Everything: Capitalism vs. the Climate* (New York: Simon & Schuster, 2015), 92.

8. Well-argued anti-growth thinking along these lines can be found in Giacomo D'Alisa, Fed- erico Demaria, and Giorgos Kallis, eds., *Degrowth: A Vocabulary for a New Era* (Abingdon and New York: Routledge, 2014); Tim Jackson, *Prosperity without Growth: Economics for a Finite Planet* (London: Earthscan, 2011); Giorgos Kallis, Christian Kerschner, and Joan Martinez-Alier, "The Economics of Degrowth," *Ecological Economics* 84 (2012): 172–180; Klein, *This Changes Everything*; T. Parrique et al., *Decoupling Debunked: Evidence and Arguments against Green Growth* (European Environmental Bureau, 2019); Peter A. Victor, *Managing without Growth: Slower by Design, Not Disaster* (Cheltenham, UK: Edward Elgar Publishing, 2019).

9. Steve Keen, "Mad, Bad and Dangerous to Know," *Real-World Economics Review* 49 (2009): 2–7.

10. See Anders Hayden, *When Green Growth Is Not Enough* (Montreal: McGill-Queen's University Press, 2014); Duncan Austin, "Greenwish: The Wishful Thinking Under- mining the Ambition of Sustainable Business," *Real-World Economics Review* 90 (2019): 47; Noel B. Verrinder et al., "Evaluative Tools in Impact Investing: Three Case Studies on the Use of Theories of Change," *African Evaluation Journal* 6, no. 2 (2018); Rob van Tulder and Laura Lucht, "Reversing Materiality: From a Reactive Matrix to a Proactive SDG Agenda," in *Innovation for Sustainability*, ed. Nancy Bocken et al. (Cham: Springer, 2019), 271–289; R. Nieuwenkamp, "Ever Heard of SDG Washing? The Urgency of SDG Due Diligence," OECD Development Matters, September 25, 2017, https://oecd -development-matters.org/2017/09/25/ever-heard-of-sdg-washing-the-urgency-of-sdg -due-diligence/, accessed May 2, 2020.

11. Lawrence M. Heim, *Killing Sustainability: Blunt Truths about Corporate Sustainability/Social Responsibility Failures and How to Avoid Them* (self-pub., 2018).

12. John Elkington, "25 Years Ago I Coined the Phrase 'Triple Bottom Line.' Here's Why It's Time to Rethink It," *Harvard Business Review* 25 (2018); GreenBiz Group and TruCost ESG, *State of Green Business 2019* (Oakland, CA, January 2019); Austin, "Greenwish"; Parrique et al., *Decoupling Debunked*.

13. Robert Jay Lifton, *The Protean Self: Human Resilience in an Age of Fragmentation* (Chicago: University of Chicago Press, 1999), 21.

14. On the inclusive, ecological self, see Arne Naess, "Self-Realization" in *Thinking like a Mountain*, ed. John Seed et al. (Gabriola Island, BC: New Society Publishers, 2007). On enlivenment, see Andreas Weber, *Enlivenment Towards a Fundamental Shift in the Concepts of Nature, Culture and Politics* (Berlin: Heinrich Böll Stiftung, 2013).

15. Donella H. Meadows et al., *The Limits to Growth: A Report for the Club of Rome's Project on the Predicament of Mankind* (New York: Universe Books, 1972).

16. Deirdre N. McCloskey, *The Rhetoric of Economics* (Madison: University of Wisconsin Press, 1998); Donald N. McCloskey, "Metaphors Economists Live By," *Social Research* (1995): 215–237.

17. Jorgen Randers et al., *Transformation Is Feasible—How to Achieve the 17 SDGs within Planetary Boundaries* (Stockholm: Stockholm Resilience Center, 2018), https://www.stockholmresilience.org/publications/artiklar/2018–10–17-transformation-is-feasible-how-to-achieve-the-sustainable-development-goals-within-planetary-boundaries.html; Jorgen Randers et al., "Achieving the 17 Sustainable Development Goals within 9 Planetary Boundaries," *Global Sustainability* 2 (2019); Paul Hawken, ed., *Drawdown: The Most Comprehensive Plan Ever Proposed to Reverse Global Warming* (New York: Penguin Books, 2017); IRENA, *Global Energy Transformation: A Roadmap to 2050 (2019 Edition)* (Abu Dhabi: International Renewable Energy Agency, 2019); Energy Transitions Commission, *Mission Possible: Reaching Net-Zero Carbon Emissions from Harder-to-Abate Sectors by Mid-Century* (2018), http://www.energy-transitions.org/sites/default/files/ETC_Mission Possible_FullReport.pdf; M. Ram et al., *Global Energy System Based on 100% Renewable Energy—Power, Heat, Transport and Desalination Sectors* (Study by Lappeenranta University of Technology and Energy Watch Group, Lappeenranta, Berlin, 2019); Arnulf Grubler et al., "A Low Energy Demand Scenario for Meeting the 1.5°C Target," *Nature Energy* 3, no. 6 (2018): 515–527.

CHAPTER 1

1. George Lakoff and Mark Johnson, *Metaphors We Live By* (Chicago: University of Chicago Press, 2003), 15; George Lakoff, *The Political Mind: A Cognitive Scientist's Guide to Your Brain and Its Politics* (New York: Penguin Books, 2009).

2. This area of generative or "conceptual metaphors" and their effects on our language and thinking is a core research topic within cognitive linguistics and broader cognitive

neurosciences. See George Lakoff, *Women, Fire, and Dangerous Things: What Categories Reveal about the Mind* (Chicago: University of Chicago Press, 2012); Drew Westen, *The Political Brain: The Role of Emotion in Deciding the Fate of the Nation* (New York: Public Affairs, 2008); Lawrence A. Shapiro, *Embodied Cognition* (New York: Routledge, 2019).

3. World Bank and Commission on Growth and Development, *The Growth Report* (2008).

4. Matthias Schmelzer, *The Hegemony of Growth: The OECD and the Making of the Economic Growth Paradigm* (Cambridge: Cambridge University Press, 2016).

5. Samuelson and Nordhaus, *Economics*, 567.

6. Martin Jacques, *When China Rules the World: The End of the Western World and the Birth of a New Global Order* (New York: Penguin Books, 2012).

7. William D. Nordhaus, *A Question of Balance: Weighing the Options on Global Warming Policies* (New Haven: Yale University Press, 2008); M. Baily and F. Comes, "Prospects for Growth: An Interview with Robert Solow," *McKinsey Quarterly* (2014); Robert J. Gordon, *The Rise and Fall of American Growth: The U.S. Standard of Living since the Civil War* (Princeton, NJ: Princeton University Press, 2016).

8. Andrew Soergel, "How Long Will the Economic Recovery Last?," *US News & World Report*, July 25, 2017, https://www.usnews.com/news/the-report/articles/2017-07-25/how-long -will-the-economic-recovery-last; my italics.

9. Clive Crook, "The World According to CSR," *Economist* 374, no. 10 (2005).

10. For instance, President Trump said of the GDP indicator, "If it has a 4 in front of it, we're happy." Tucker Higgins, Jacob Pramuk, and Mike Calia, "Trump Triumphant," *CNBC*, July 27, 2018, https://www.cnbc.com/2018/07/27/trump-were-going-to-go-a-lot-higher-than -4point1percent-gdp-number.html.

11. "Global Publics More Upbeat about the Economy," Pew Research Center, June 5, 2017, https://www.pewresearch.org/global/2017/06/05/global-publics-more-upbeat-about-the -economy/, accessed July 2, 2020.

12. "Most Important Problem," Gallup, https://news.gallup.com/poll/1675/most-important -problem.aspx, accessed April 10, 2020. But in 2019, US citizens mentioned "the govern- ment" as the top problem, far outweighing economic concerns.

13. Robert Ayres, "Limits to the Growth Paradigm," *Ecological Economics* 19, no. 2 (1996): 117–134; Robert Ayres, *Turning Point: End of the Growth Paradigm* (New York: Routledge, 2014); Schmelzer, *The Hegemony of Growth*; Giorgos Kallis, *Degrowth* (Newcastle upon Tyne: Agenda Publishing, 2018).

14. R. J. Douthwaite, *The Growth Illusion: How Economic Growth Has Enriched the Few, Impover- ished the Many, and Endangered the Planet* (Bideford, Devon: Green Books, 1992); Herman E. Daly, *Beyond Growth: The Economics of Sustainable Development* (Boston: Beacon Press, 1997); Jackson, *Prosperity without Growth*; Richard Heinberg, *The End of Growth: Adapting to Our New Economic Reality* (Gabriola Island, BC: New Society Publishers, 2011); Victor, *Managing without Growth*.

15. Meadows et al., *The Limits to Growth.*

16. Graham Turner, *Is Global Collapse Imminent? An Updated Comparison of The Limits to Growth with Historical Data* (MSSI Research Paper No. 4, Melbourne Sustainable Society Institute, The University of Melbourne, 2014).

17. Henry Wallich, "To Grow or Not to Grow," *Newsweek*, March 13, 1972.

18. In a 1983 speech Reagan said, "There are no such things as limits to growth, because there are no limits on the human capacity for intelligence, imagination, and wonder": https://reaganlibrary.archives.gov/archives/speeches/1983/92083c.htm. He found this topic important enough to repeat again in his inaugural speech in 1985: "We believed then and now there are no limits to growth and human progress when men and women are free to follow their dreams." See http://www.bartleby.com/124/pres62.html, accessed May 2, 2020.

19. World Wildlife Fund, *Living Planet Report—2018: Aiming Higher* (Gland, Switzerland, 2018).

20. Will Steffen et al., "The Trajectory of the Anthropocene: The Great Acceleration," *Anthropocene Review* 2, no. 1 (April 2015): 81–98.

21. Martin R. Stuchtey, Per-Anders Enkvist, and Klaus Zumwinkel, *A Good Disruption: Redefining Growth in the Twenty-First Century* (London: Bloomsbury, 2016): 248–252.

22. Laurence Fink, "Our Gambling Culture," McKinsey.com, April 1, 2015, https://www.mckinsey.com/business-functions/strategy-and-corporate-finance/our-insights/our-gambling-culture.

23. Jorgen Randers, 2018, personal comm. See also Graeme P. Maxton, Jorgen Randers, and David T. Suzuki, *Reinventing Prosperity: Managing Economic Growth to Reduce Unemployment, Inequality, and Climate Change: A Report to the Club of Rome* (Vancouver and Berkeley: Greystone Books, 2016).

24. Herman Daly calls this "preanalytic visions." Herman E. Daly and Joshua C. Farley, *Ecological Economics: Principles and Applications* (Washington, DC: Island Press, 2010), ch. 2.

25. Daniel Kahneman, *Thinking, Fast and Slow* (New York: Farrar, Straus and Giroux, 2013); Ziva Kunda, "The Case for Motivated Reasoning," *Psychological Bulletin* 108, no. 3 (1990): 480; Mark R. Leary and June Price Tangney, *Handbook of Self and Identity* (New York: Guilford Press, 2011).

26. Lyle Scruggs and Salil Benegal, "Declining Public Concern about Climate Change: Can We Blame the Great Recession?," *Global Environmental Change* 22, no. 2 (May 2012): 505–515.

27. Degrowth proponents are unclear on whether degrowth means GDP decline (i.e., an economic contraction) or simply abandoning the concept itself (i.e., stop making national accounts for GDP). Sometimes degrowth is not a contraction; other times a GDP decline is inevitable. See Kallis, *Degrowth* chs. 1, 6.

28. World Bank, *Inclusive Green Growth: The Pathway to Sustainable Development* (Washington, DC: World Bank, 2012), http://documents.worldbank.org/curated/en/2012/01/16283976/inclusive-green-growth-pathway-sustainable-development.

29. European Commission, "Managing Resources," http://ec.europa.eu/environment/basics/green-economy/resources/index_en.htm, accessed April 10, 2020.

30. Timan Santarius, *Green Growth Unravelled: How Rebound Effects Baffle Sustainability Targets When the Economy Keeps Growing* (Heinrich Böll Foundation, Wuppertal Institute for Climate, Environment and Energy, 2012); Tilman Santarius, Hans Jokob Walnum, and Carlo All, eds., *Rethinking Climate and Energy Policies* (New York: Springer, 2016); Hans Walnum, Carlo Aall, and Søren Løkke, "Can Rebound Effects Explain Why Sustainable Mobility Has Not Been Achieved?," *Sustainability* 6, no. 12 (2014): 9510–9537.

31. Hayden, *When Green Growth Is Not Enough*; Oliver Taherzadeh and Benedict Probst, "Five Reasons 'Green Growth' Won't Save the Planet," *The Conversation*, May 20, 2019.

32. Jackson, *Prosperity without Growth*; Victor, *Managing without Growth*; Tim Jackson, "The Post-Growth Challenge: Secular Stagnation, Inequality and the Limits to Growth," *Ecological Economics* 156 (February 2019): 236–246; Stefan Drews and Miklós Antal, "Degrowth: A 'Missile Word' That Backfires?," *Ecological Economics* 126 (June 2016): 182–187.

33. Fritjof Capra and Hazel Henderson, "Qualitative Growth: A Conceptual Framework," in *From Capitalistic to Humanistic Business*, ed. Michael Pirson et al. (London: Palgrave Macmillan UK, 2014), 35–47.

34. Gro Harlem Brundtland, *Our Common Future* (New York: United Nations, 1987).

35. Kate Raworth, *Doughnut Economics: Seven Ways to Think like a 21st Century Economist* (White River Junction, VT: Chelsea Green Publishing, 2017), ch. 7. Van den Bergh recommends the concept of "agrowth," similar to Raworth's agnostic principle: Jeroen van den Bergh, "Agrowth Instead of Anti-and Pro-Growth: Less Polarization, More Support for Sustainability/Climate Policies," *Journal of Population and Sustainability* 3, no. 1 (2018): 53–73.

36. Van den Bergh, "Agrowth Instead of Anti-and Pro-Growth."

37. The theory on this loop can be found in, for instance, Gareth Morgan, *Imaginization: New Mindsets for Seeing, Organizing and Managing* (San Francisco: Sage Publications, 1997), appendix A; James Hillman, *Kinds of Power* (New York: Penguin, 1997), part I.

38. Peter L. Berger and Thomas Luckmann, *The Social Construction of Reality: A Treatise in the Sociology of Knowledge* (New York: Anchor Books, 1990).

39. Gareth Morgan, *Images of Organization* (New York: Sage Publications, 2016).

40. Hillman, *Kinds of Power*, 64.

41. Hawken, *Drawdown*.

42. Three trillion trees: Thomas W. Crowther, Henry B. Glick, Kristofer Covey, et al., "Mapping Tree Density at a Global Scale," *Nature* 525, no. 7568 (2015): 201–205. Tree overpopulation: William McDonough, *The Upcycle* (New York: North Point Press, 2013).

43. James Hillman, *Re-Visioning Psychology* (New York: HarperPerennial, 1992), xiii.

44. In archetypal psychology, "strict father" and "nurturing mother" are metaphors, not gender stereotypes. Hence, a woman may speak or behave in a "strict father" pattern and a man as "nurturing mother." We keep the gendered metaphors as they open the imagination to archaic myths and patterns, not as literal depictions. See Hillman, *Re-Visioning Psychology*.

45. Since the first industrial revolution got going around 1750, this imperative of growth has guided the development of technology and capitalist power. See David S. Landes, *The Unbound Prometheus: Technological Change and Industrial Development in Western Europe from 1750 to the Present*, 2nd ed. (Cambridge: Cambridge University Press, 2003); John S. Dryzek, *The Politics of the Earth: Environmental Discourses*, 3rd ed. (Oxford: Oxford University Press, 2013), ch. 3.

46. Walter Logeman, *Archetypes of Cyberspace: An Essay* (Amazon, 2014); Ginette Paris, *Pagan Grace: Dionysus, Hermes and Goddess Memory* (Putnam, CT, Spring Publications, 1990), part II.

47. James Hillman, *Archetypal Psychology* (Putnam, CT: Spring Publications, 2004).

48. Jung, *The Collected Works*, vol. 10, iv, §490.

49. Clive Spash, "Better Growth, Helping the Paris COP-out? Fallacies and Omissions of the New Climate Economy Report" *SRE—Discussion Papers* (April 2014). Schmelzer, *The Hegemony of Growth*.

50. Andrew Ortony, ed., *Metaphor and Thought* (Cambridge and New York: Cambridge University Press, 1993).

51. Alex Steffen, "Announcing Our New Project, The Heroic Future," AlexSteffen.com (blog), February 7, 2016, http://www.alexsteffen.com/announcing_heroic_future.

52. Klein, *This Changes Everything*.

53. Edward Wolff, "Household Wealth Trends in the United States, 1962–2013" (Cambridge, MA: National Bureau of Economic Research, December 2014); Jared Bernstein, "Yes, Stocks Are Up. But 80 Percent of the Value Is Held by the Richest 10 Percent," *Washington Post*, March 2, 2017.

54. Per Espen Stoknes, *Money & Soul: The Psychology of Money and the Transformation of Capitalism* (Totnes, Devon: Green Books, 2009), chs. 1–2.

55. Robert L. Heilbroner, *The Worldly Philosophers: The Lives, Times, and Ideas of the Great Economic Thinkers* (New York: Simon & Schuster, 1999), ch. 11.

56. McCloskey, "Metaphors Economists Live By"; Stoknes, *Money & Soul*.

57. Stoknes, *Money & Soul*, ch. 6.

58. Maeve Cohen, "Post-Crash Economics: Have We Learnt Nothing?," *Nature* 561 (2018); "The World Has Not Learned the Lessons of the Financial Crisis," *Economist*, September 6, 2018.

CHAPTER 2

1. John Elkington, *Green Swans* (New York: Fast Company, 2020); Paul Hawken, Amory B. Lovins, and L. Hunter Lovins, *Natural Capitalism: The Next Industrial Revolution* (London: Earthscan, 2005); Paul Hawken, *Regeneration: Ending the Climate Crisis in One Generation* (New York: Penguin Books, forthcoming).

2. This and the following quotations are from Tristram Stuart, personal communication with the author.

3. *Invest in Women, Power Communities: 2019–2023 Strategy* (Solar Sister, February 1, 2019), https://solarsister.org/wp-content/uploads/2019/01/Solar-Sister-2019-2023-web.pdf.

4. This and the following quotations are from Neha Misra, personal communication with the author.

5. Jonathan Woetzel, "The Power of Parity: How Advancing Women's Equality Can Add \$12 Trillion to Global Growth," working paper, 2015; Victoria Bateman, *The Sex Factor: How Women Made the West Rich* (Cambridge: Polity Press, 2019).

6. My talk on the sun-rich future is available on TED.com: "The Sunny Future of Regenerative Economies," TEDxGöteborg, October 2017, https://www.ted.com/talks/per_espen_stoknes_the_sunny_future_of_regenerative_economies_may_2018.

7. Paul Vitello, "Ray Anderson, Business Man Turned Environmentalist, Dies at 77," *New York Times*, August 10, 2011, https://www.nytimes.com/2011/08/11/business/ray-anderson-a-carpet-innovator-dies-at-77.html.

8. Ray Anderson and John A. Lanier, *Mid-Course Correction Revisited: The Story and Legacy of a Radical Industrialist and his Quest for Authentic Change* (White River Junction, VT: Chelsea Green Publishing, 2019).

9. Anderson and Lanier, *Mid-Course Correction Revisited*.

10. Anderson and Lanier, *Mid-Course Correction Revisited*.

11. "The buildings and building construction sectors combined are responsible for 36% of global final energy consumption and nearly 40% of total direct and indirect CO_2 emissions." International Energy Agency IEA (2019), based on 2017 data. Source: https://www.iea.org/topics/energyefficiency/buildings/, accessed October 10, 2019.

12. For more information, see www.powerhouse.no/en.

13. Power generation is not included in this list as a separate sector, because all the power generated is sold from energy plants to players in these sectors where end-users are consuming it. If we can electrify and reduce the energy consumption at the end-user side, this cascades back to lower need for production of power and primary fossil energy. Source for end-user sector breakdown of emissions is from IPCC, figure 1.3, in IPCC, ed., *Climate Change 2014: Mitigation of Climate Change: Working Group III Contribution to the Fifth Assessment Report of the Intergovernmental Panel on Climate Change* (New York: Cambridge University Press, 2014), 123.

14. The prize was given in July 2019, as reported by Bioenergy Insight, "AD and biogas industry celebrates its 2019 champions" (July 4, 2019), https://www.bioenergy-news.com/news/ad-and-biogas-industry-celebrates-its-2019-champions/.

15. Kari-Anne Lyng and Andreas Brekke, "Environmental Life Cycle Assessment of Biogas as a Fuel for Transport Compared with Alternative Fuels," *Energies* 12, no. 3 (2019): 532; Qingyao He et al., "B.E.E.F: A Sustainable Process Concerning Negative CO_2 Emission and Profit Increase of Anaerobic Digestion," *ACS Sustainable Chemistry & Engineering* 7, no. 2 (2019): 2276–2284.

16. See the further reading section at the end of this volume, for a brief guide to further resources.

17. Per Espen Stoknes, *What We Think about When We Try Not to Think about Global Warming: Toward a New Psychology of Climate Action* (White River Junction, VT: Chelsea Green Publishing, 2015), ch. 10.

18. Julie Allan, Gerard Fairtlough, and Barbara Heinzen, *The Power of the Tale: Using Narratives for Organisational Success* (Chichester and New York: Wiley, 2002); David C. Korten, *Change the Story, Change the Future: A Living Economy for a Living Earth* (Oakland, CA: Berrett-Koehler, 2015); Dominic McAfee et al., "Everyone Loves a Success Story: Optimism Inspires Conservation Engagement," *BioScience* 69, no. 4 (2019): 274–281; Stoknes, *What We Think about When We Try Not to Think about Global Warming*, ch. 12.

CHAPTER 3

1. Winston Churchill, "House of Commons Rebuilding" (HC Deb vol. 393 cc403-73, October 28, 1943), https://api.parliament.uk/historic-hansard/commons/1943/oct/28/house-of-commons-rebuilding.

2. Canadian philosopher Marshall McLuhan described how new technologies (such as alphabets, printing presses, and even speech) had powerful effects on our cognition, which in turn affects social organization. With too little room to delve into this topic, I only want to mention the tradition of work building on Marshall McLuhan, *The Gutenberg Galaxy: The Making of Typographic Man* (Toronto: University of Toronto Press, 2011); Marshall McLuhan and Bruce R. Powers, *The Global Village: Transformations in World Life and Media in the 21th Century* (New York: Oxford University Press, 1992).

3. Sherry Turkle, "How Computers Change the Way We Think," *Chronicle of Higher Education* 50, no. 21 (2004): B26.

4. Daniel Kahneman, Jack L Knetsch, and Richard H. Thaler, "Anomalies: The Endowment Effect, Loss Aversion, and Status Quo Bias," *Journal of Economic Perspectives* 5, no. 1 (1991): 193–206; Elke U. Weber, "Breaking Cognitive Barriers to a Sustainable Future," *Nature Human Behaviour* 1, no. 1 (2017): 1–2.

5. IEA publishes the World Energy Outlook annually, at least since 2000, with such scenarios. For a good overview of energy scenarios to 2050, see Dawud Ansari, Franziska Holz, and

Hashem Al-Kuhlani, "Energy Outlooks Compared: Global and Regional Insights" (DIW Berlin: Politikberatung kompakt, 2019).

6.   In the following, I present my interpretation of the innovation wave theory, inspired in particular by Perez and Weizsäcker: Carlota Perez, *Technological Revolutions and Financial Capital: The Dynamics of Bubbles and Golden Ages* (Cheltenham: Elgar, 2003); Ernst U. von Weizsäcker, *Factor Five: Transforming the Global Economy through 80% Improvements in Resource Productivity* (London: Earthscan, 2009).

7.   In his 1942 classic, Schumpeter described the "gale of creative destruction" as a "process of industrial mutation that incessantly revolutionizes the economic structure from within, incessantly destroying the old one, incessantly creating a new one." Joseph A. Schumpeter, *Capitalism, Socialism and Democracy* (New York: HarperPerennial, 1942).

8.   Joseph Schumpeter, *Business Cycles*, vol. 1 (New York: McGraw-Hill, 1939); Nikolai D. Kondratieff, "The Long Waves in Economic Life," *Review (Fernand Braudel Center)* (1979): 519–562; Perez, *Technological Revolutions and Financial Capital*.

9.   Carlota Perez calls them techno-economic paradigm shifts, albeit in a different sense of that word than in scientific paradigm shift, as in Kuhn's classic. Thomas S. Kuhn, *The Structure of Scientific Revolutions*, 3rd ed. (Chicago: University of Chicago Press, 1996).

10. Mariana Mazzucato et al., "Innovation as Growth Policy," *The Triple Challenge for Europe: Economic Development, Climate Change, and Governance* (Oxford: Oxford University Press, 2015), 229–264.

11. Source for figure 3.1: graphics by the author. Data sources: Nicholas Felton, "Consumption Spreads Faster Today," *New York Times*, February 10, 2008; "U.S. Smartphone Penetration Surpassed 80 Percent in 2016," Comscore, Inc., https://www.comscore.com/Insights/Blog/US-Smartphone-Penetration-Surpassed-80-Percent-in-2016, accessed August 12, 2019.

12. Jeff Desjardins, "The Largest Companies by Market Cap Over 15 Years," *VisualCapitalist*, August 12, 2016, https://www.visualcapitalist.com/chart-largest-companies-market-cap -15-years/; Statista and Erin Duffin, "Biggest Companies in the World 2018, Source Forbes," Statista, August 12, 2019, https://www.statista.com/statistics/263264/top -companies-in-the-world-by-market-value/, accessed Dec. 2, 2019.

13. For global LCOE of solar 2010–2018 averages, see UNEP, *Emissions Gap Report* (Nairobi: United Nations Environment Program, 2019), 47. For solar module prices 2010–2019, see IRENA, *Renewable Power Generation Costs in 2019* (Abu Dhabi, 2020).

14. World Economic Forum, *Transformation of the Global Energy System* (Cologny, Switzerland: World Economic Forum, March 2, 2018); DNV GL, *Energy Transition Outlook 2019* (Oslo: DNV GL, October 2019); John Mathews, *Global Green Shift: When Ceres Meets Gaia* (London: Anthem Press, 2017).

15. Wolfgang J. M. Drechsler, Rainer Kattel, and Erik S. Reinert, eds., *Techno-Economic Paradigms: Essays in Honour of Carlota Perez* (London: Anthem Press, 2011); Perez, *Technological Revolutions and Financial Capital*.

16. World Business Council for Sustainable Development, *WBCSD—Vision 2050* (September 4, 2012), https://www.wbcsd.org/Overview/About-us/Vision2050/Resources/Vision2050-Road-Map.

17. Neil J. Smelser and Richard Swedberg, *The Handbook of Economic Sociology* (Princeton, NJ: Princeton University Press, 2010).

18. M. De Wit et al., *The Circularity Gap Report: An Analysis of the Circular State of the Global Economy* (Amsterdam: *Circle Economy*, 2019).

19. IRP et al., *Global Resources Outlook 2019: Natural Resources for the Future We Want* (A Report of the International Resource Panel, United Nations Environment Programme, Nairobi, Kenya, 2019), 56; Andrew McAfee, *More from Less: The Surprising Story of How We Learned to Prosper Using Fewer Resources—and What Happens Next* (London: Simon & Schuster, 2019), ch. 5.

20. In this book, resource *productivity* means value creation (or value added) divided by resource use in physical units, and can be expressed in $/ton or $/kWh, while resource *efficiency* refers to the factor of improvement in benefit per resource unit, as in W/lumens of light or miles/gallon. For instance, changing from using 100W incandescent lightbulb to 10W LED to get the same 1600 lumens lighting. This represents 90 percent improvement in resource efficiency, or a factor 10 less energy use at end user. Better resource efficiency will as a rule improve resource productivity, but since other factors also influence value creation (such as market prices, competition, capital and labor costs), there is rarely a 1:1 relationship between them. More on the energy efficiency potential here: Amory B. Lovins, "How Big Is the Energy Efficiency Resource?," *Environmental Research Letters* 13, no. 9 (2018): 1–17.

21. Intensity is the mathematical inverse of productivity. The two measures of carbon productivity and carbon intensities, for instance, track the same relationship between value creation and carbon emissions, but intensity usually trends downward toward zero ($tCO_2/\$$) and productivity usually trends upward over time ($\$/tCO_2$).

22. Shared taxis or "taxi-buses" are real-time ride-sharing cars or minivans with multiple passengers on similar routes, booked on short notice via apps so wait times are short.

23. The Low-Energy-Demand (LED) scenario project has identified 21 potentially disruptive end-user innovations. See A. Grubler et al., "A Low Energy Demand Scenario for Meeting the 1.5°C Target"; Charlie Wilson, "Transforming Energy Demand to Meet the 1.5°C Target and Sustainable Development Goals without Negative Emission Technologies" (IEA, November 2018); Charlie Wilson et al., "The Potential Contribution of Disruptive Low-Carbon Innovations to 1.5°C Climate Mitigation," *Energy Efficiency* 12, no. 2 (February 2019): 423–440.

24. Estimate for solar modules is 26–28 percent cost reduction in $ per watt, according to Jenny Chase and Shahan Zachary, "Solar Panel Prices Continue Falling Quicker Than Expected," *CleanTechnica*, February 11, 2018, https://cleantechnica.com/2018/02/11/solar-panel-prices-continue-falling-quicker-expected-cleantechnica-exclusive/.

25. AllianceBernstein, "What's the Long-Term Oil Outlook? Watch How the Majors Invest," https://www.alliancebernstein.com/library/whats-the-long-term-oil-outlook-watch-how-the-majors-invest.htm, accessed April 17, 2020.

26. New Energy Finance Bloomberg, *New Energy Outlook, 2019.*

27. Kazuhisa Miyamoto, *Renewable Biological Systems for Alternative Sustainable Energy Production*, vol. 128 (Food & Agriculture Org., 1997).

28. On perovskite, see for instance "Perovskite Solar Cells," NREL, https://www.nrel.gov/pv/perovskite-solar-cells.html, accessed January 1, 2020.

29. While costs have fallen between 20–30 percent in $/kWh per year, the learning rates (cost decline per doubling of volume) has been around 18 percent. New Energy Finance Bloomberg and Logan Goldie-Scot, "A Behind the Scenes Take on Lithium-Ion Battery Prices," *BloombergNEF*, March 5, 2019, https://about.bnef.com/blog/behind-scenes-take-lithium-ion-battery-prices/.

30. In 2019, 2.3 million cars sold had electric plugs. Source: EV-Volumes.com, http://www.ev-volumes.com/country/total-world-plug-in-vehicle-volumes/, accessed April 10, 2020.

31. Mark Lewis, "Wells, Wires and Wheels: EROCI and the Tough Road Ahead for Oil," BNP Paribas, August 2019, https://investors-corner.bnpparibas-am.com/investment-themes/sri/petrol-eroci-petroleum-age/; "The Death of the Internal Combustion Engine—Electric Cars," *Economist*, August 12, 2017, https://www.economist.com/leaders/2017/08/12/the-death-of-the-internal-combustion-engine.

32. 8.4 gigatons of materials are cycled input, versus 84.4 gigatons coming from extracted resources in 2015. Source: De Wit et al., *The Circularity Gap Report.*

33. See EU EC, *Circular Economy Action Plan* (Brussels: EC, 2020), https://ec.europa.eu/environment/circular-economy/pdf/new_circular_economy_action_plan.pdf; EU EC, *Circular Economy Factsheet* (Brussels: EC, 2016), https://ec.europa.eu/commission/sites/beta-political/files/circular-economy-factsheet-general_en.pdf; Perez et al., "Changing Gear in R&I" (Brussels: EC, 2016), https://op.europa.eu/s/n3my (all accessed 1.Apr.2020).

34. Vaclav Smil, *Making the Modern World: Materials and Dematerialization* (Chichester, UK: Wiley, 2014).

35. See, for instance, Blue Planet Ltd's approach (http://www.blueplanet-ltd.com/#products) and discussion in Ramses Kools, "CO_2 Sequestration and Utilization in Cement-Based Materials," master's thesis, University of Amsterdam, 2018.

36. The old mall is called CC-Vest and the owner and developer is Mustad Eiendom, www.mustadeiendom.no. Figures were disclosed at the Urban-Future.org conference in Oslo, May 23, 2019.

37. Source: "Apple Expands Global Recycling Program," Apple (press release), April 4, 2019, https://www.apple.com/newsroom/2019/04/apple-expands-global-recycling-programs/.

38. Energy Transitions Commission, *Mission Possible*, 16, 99.

39. Energy Transitions Commission, *Mission Possible*, 99; *Completing the Picture: How the Circular Economy Tackles Climate Change* (Ellen MacArthur Foundation, 2019), www .ellenmacarthurfoundation.org/publications.

40. De Wit et al., *The Circularity Gap Report*, 17.

41. Stuchtey et al., *A Good Disruption*; Maria Antikainen, Teuvo Uusitaloa, and Päivi Kivikytö-Reponena, "Digitalisation as an Enabler of Circular Economy," *Procedia CIRP* 73 (2018): 45–49.

42. For a more in-depth discussion of EROI or NER, see Adam R. Brandt et al., "Energy Return on Investment (EROI) for Forty Global Oilfields Using a Detailed Engineering-Based Model of Oil Production," *PLoS ONE* 10, no. 12 (2015); Nathan Gagnon, Charles A. S. Hall, and Lysle Brinker, "A Preliminary Investigation of Energy Return on Energy Investment for Global Oil and Gas Production," *Energies* 2, no. 3 (2009): 490–503.

43. Devin Moeller and David Murphy, "Net Energy Analysis of Gas Production from the Marcellus Shale," *BioPhysical Economics and Resource Quality* 1, no. 1 (August 2016): 12.

44. See Paul E. Brockway et al., "Estimation of Global Final-Stage Energy-Return-on-Investment for Fossil Fuels with Comparison to Renewable Energy Sources," *Nature Energy* 4, no. 7 (2019): 612–621; Cutler J. Cleveland, "Net Energy from the Extraction of Oil and Gas in the United States," *Energy* 30, no. 5 (2005): 769–782; Gagnon, Hall, and Brinker, "A Preliminary Investigation of Energy Return."

45. Recent (2018) studies show a high global social cost of carbon values, a median of US$417 per ton of CO_2 or around US$200 per barrel. Katharine Ricke et al., "Country-Level Social Cost of Carbon," *Nature Climate Change* 8, no. 10 (October 2018): 895–900.

46. See "Conventional Discoveries Have Fallen to Lowest Levels in 70 Years," https://www .businesswire.com/news/home/20191001006010/en/, accessed March 1, 2020.

47. Josh Gabbatiss, "China's Emissions 'Could Peak 10 Years Earlier than Paris Climate Pledge,'" Carbon Brief, July 29, 2019, https://www.carbonbrief.org/chinas-emissions-could -peak-10-years-earlier-than-paris-climate-pledge. See also McKinsey's analysis: "Is Peak Oil Demand in Sight?," McKinsey, June 2016, http://www.mckinsey.com/industries/ oil-and-gas/our-insights/is-peak-oil-demand-in-sight.

48. Billy Nauman, "Sharp Rise in Number of Investors Dumping Fossil Fuel Stocks," *Financial Times*, September 9, 2019, https://www.ft.com/content/4dec2ce0-d0fc-11e9–99a4-b5 ded7a7fe3f; Christina Atanasova and Eduardo S. Schwartz, "Stranded Fossil Fuel Reserves and Firm Value" (National Bureau of Economic Research, 2019).

49. For instance, IRENA makes a scenario study to 2050 named ReMap that delivers on keeping global warming well below 2°C. In the scenario "fossil fuel subsidies [are] USD 15 trillion below what would have occurred in the Reference Case by 2050, and [result] in a net reduction of USD 10 trillion when including the increased support needed for renewables in the REmap Case. In total the savings from avoided subsidies and reduced environmental and health damages are about three to seven times larger than the additional energy system

costs. In monetary terms, total savings resulting from the REmap Case could amount to between USD 65 trillion and USD 160 trillion over the period to 2050. Viewed differently, *for every USD 1 spent, the payoff would be between USD 3 and USD 7*" (my italics). For more future-looking analyses of the coming shift, see: Jeremy Rifkin, *The Zero Marginal Cost Society* (New York: Palgrave Macmillan, 2015); Tony Seba, *Clean Disruption of Energy and Transportation: How Silicon Valley Will Make Oil, Nuclear, Natural Gas, Coal, Electric Utilities and Conventional Cars Obsolete by 2030* (self-pub., 2014); Global Commission on the Economy and Climate, *Unlocking the Inclusive Growth Story of the 21st Century, NCER* (Washington, DC: The Global Commission on the Economy and Climate, 2018); Lewis, "Wells, Wires and Wheels"; IRENA, *Global Energy Transformation*.

50.   Energy Transitions Commission, *Mission Possible*.

51.   Despite political pledges to tackle global heating, governments' global fossil-fuel support, inclusive of negative externality costs is US$5.2 trillion, or 6.5 percent of world GDP. See David Coady et al., *Global Fossil Fuel Subsidies Remain Large: An Update Based on Country-Level Estimates* (International Monetary Fund, 2019).

52.   Lewis, "Wells, Wires and Wheels."

53.   To use the phrase from Martin Stuchtey, former head of McKinsey Center for Business and Environment, in Stuchtey, Enkvist, and Zumwinkel, *A Good Disruption*.

CHAPTER 4

1.   Glen Slater, "Disney's Wall-E, a Movie Review," *Spring: A Journal of Archetype and Culture* 80 (2009).

2.   Olivier De Schutter, "The Political Economy of Food Systems Reform," *European Review of Agricultural Economics* 44, no. 4 (September 2017): 705–731.

3.   Grubler et al., "A Low Energy Demand Scenario for Meeting the 1.5°C Target"; DNV GL, "Energy Transition Outlook—A Global and Regional Forecast to 2050" (DNV GL, October 2019), https://eto.dnvgl.com/2019/; IRENA, *Global Energy Transformation*; Ram et al., *Global Energy System Based on 100% Renewable Energy*.

4.   Source: IEA (2019), https://www.iea.org/tcep/buildings/lighting/, accessed September 1, 2019.

5.   Especially with "integrative designs." See Lovins, "How Big Is the Energy Efficiency Resource?," 5; Grubler et al., "A Low Energy Demand Scenario for Meeting the 1.5°C Target"; Global Commission on the Economy and Climate, *Unlocking the Inclusive Growth Story of the 21st Century, NCER*, 49–50.

6.   Source: IEA (2019), https://www.iea.org/tcep/buildings/lighting/, accessed July 10, 2019.

7.   "The total energy required for bottled water will typically range from 5.6 to 10.2 MJ/l. In comparison, producing tap water typically requires about 0.005 MJ/l for treatment and distribution." P. H. Gleick and H. S. Cooley, "Energy Implications of Bottled Water," *Environmental Research Letters* 4, no. 1 (January 2009): 1–6. See also Marianna Garfí et al., "Life

Cycle Assessment of Drinking Water: Comparing Conventional Water Treatment, Reverse Osmosis and Mineral Water in Glass and Plastic Bottles," *Journal of Cleaner Production* 137 (2016): 997–1003.

8. Ryan Christopher Graydon et al., "Bottled Water versus Tap Water: Risk Perceptions and Drinking Water Choices at the University of South Florida," *International Journal of Sustainability in Higher Education* (June 20, 2019), IJSHE-01-2019-0003; Erik D. Olson, Diane Poling, and Gina Solomon, *Bottled Water: Pure Drink or Pure Hype?* (National Resources Defense Council, 1999); Garfí et al., "Life Cycle Assessment of Drinking Water."

9. Thayla T. Sousa-Zomer and Paulo A. Cauchick Miguel, "Sustainable Business Models as an Innovation Strategy in the Water Sector: An Empirical Investigation of a Sustainable Product-Service System," *Journal of Cleaner Production* 171 (January 2018): S119–129.

10. Sandra Laville and Matthew Taylor, "A Million Bottles a Minute: World's Plastic Binge 'as Dangerous as Climate Change,'" *Guardian*, June 28, 2017.

11. Gleick and Cooley, "Energy Implications of Bottled Water," 4.

12. Dave Cheshire, *Building Revolutions: Applying the Circular Economy to the Built Environment* (Newcastle: RIBA Publishing, 2016); Herbert Girardet, *Creating Regenerative Cities* (Abingdon, Oxon: Routledge, 2015).

13. Sources for figure 4.1: Corrado Nizzola, "Modellierung und Verbrauchsoptimierung von ottomotorischen Antriebskonzepten," PhD dissertation (ETH Zurich, 2000); Amory Lovins, *Reinventing Fire: Bold Business Solutions for the New Energy Era* (White River Junction, VT: Chelsea Green, 2013); Stefan Heck, Matt Rogers, and Paul Carroll, *Resource Revolution: How to Capture the Biggest Business Opportunity in a Century* (New York: New Harvest Publishing, 2014); E. MacArthur, K. Zumwinkel, and M. R. Stuchtey, *Growth Within: A Circular Economy Vision for a Competitive Europe* (Ellen MacArthur Foundation, 2015); US Department of Energy, "FOTW #1044, August 27, 2018: 12–30% of Energy Put into a Conventional Car is Used to Move the Car Down the Road," Energy.gov, August 27, 2019.

14. Stuchtey, Enkvist, and Zumwinkel, *A Good Disruption*, loc. 2087.

15. MacArthur, Zumwinkel, and Stuchtey, *Growth Within*, 19.

16. Data sources: Deutsche Bank, Wood MacKenzie, as referenced in Lewis, "Wells, Wires and Wheels: EROCI and the Tough Road Ahead for Oil, BNP Paribas," 17.

17. 10.2 percent according to IPCC, *Climate Change 2014*, 123, figure 1.3b.

18. According to IEA: "In fact, SUVs were responsible for all of the 3.3 million barrels a day growth in oil demand from passenger cars between 2010 and 2018, while oil use from other type of cars (excluding SUVs) declined slightly." Laura Cozzi and Apostolos Petropoulos, "Growing Preference for SUVs Challenges Emissions Reductions in Passenger Car Market," International Energy Agency, October 15, 2019, https://www.iea.org/commentaries/growing-preference-for-suvs-challenges-emissions-reductions-in-passenger-car-market.

19. Helsinki is currently introducing such an integrated mobility-as-a-service offer, called WHIM: https://www.helsinkismart.fi/portfolio-items/whim/.

20. Seba, *Clean Disruption of Energy and Transportation*; James Arbib and Tony Seba, "Rethinking Transportation 2020–2030," A RethinkX Sector Disruption Report (May 2017); Frances Sprei, "Disrupting Mobility," *Energy Research & Social Science* 37 (March 2018): 238–242; Wilson et al., "The Potential Contribution of Disruptive Low-Carbon Innovations to 1.5°C Climate Mitigation."

21. A. Shepon et al., "Energy and Protein Feed-to-Food Conversion Efficiencies in the US and Potential Food Security Gains from Dietary Changes," *Environmental Research Letters* 11, no. 10 (October 1, 2016): 105002.

22. Peter Alexander et al., "Losses, Inefficiencies and Waste in the Global Food System," *Agricultural Systems* 153 (May 2017): 195–196.

23. MacArthur, Zumwinkel, and Stuchtey, *Growth Within*, 19.

24. MacArthur, Zumwinkel, and Stuchtey, *Growth Within*, 19.

25. Cheryl D. Fryar, Margaret D. Carroll, and Cynthia L. Ogden, "Prevalence of Overweight, Obesity, and Severe Obesity among Adults Aged 20 and Over: United States, 1960–1962 through 2015–2016," National Center for Health Statistics, September 2018, https://www.cdc.gov/nchs/data/hestat/obesity_adult_15_16/obesity_adult_15_16.pdf.

26. Walter Willett et al., "Food in the Anthropocene: The EAT–Lancet Commission on Healthy Diets from Sustainable Food Systems," *Lancet* 393, no. 10170 (2019): 447–492.

27. IRP et al., "Global Resources Outlook 2019"; Global Footprint Network, "National Footprint Accounts" (2019), https://data.footprintnetwork.org/.

28. See the World Wildlife Fund for 2019 results: "July 29: Earth Overshoot Day 2019 Is Earliest Ever," July 29, 2019, https://wwf.panda.org/?350491/Earth-Overshoot-Day-2019. For footprint data for all countries, see GFN: http://data.footprintnetwork.org.

29. Worldwatch Institute published a long series of annual State of the World reports. See, for instance, the highly relevant Worldwatch Institute et al., eds., *State of the World 2013: Is Sustainability Still Possible?* (Washington, DC: Island Press, 2013).

30. Robert M. Solow, "A Contribution to the Theory of Economic Growth," *Quarterly Journal of Economics* 70, no. 1 (1956): 65–94.

31. Solow, "A Contribution to the Theory of Economic Growth," 67.

32. Solow, "A Contribution to the Theory of Economic Growth," 67.

33. Stoknes, *Money & Soul*; Pushpam Kumar, *The Economics of Ecosystems and Biodiversity: Ecological and Economic Foundations* (New York: Routledge, 2012).

34. Steffen et al., "The Trajectory of the Anthropocene."

35. Taking Solow in good faith, it is possible to read the formulation as a qualifier: it only applies "as long as" there is enough land to keep hacking away. Which is no longer the case, and therefore the Solow growth theory—according to Solow himself—no longer applies.

36. For a discussion on the role of energy in Solowian growth, see Robert Ayres and Vlasios Voudouris, "The Economic Growth Enigma: Capital, Labour and Useful Energy?," *Energy Policy* 64 (January 2014): 16–28. A more general take on growth accounting: Jane Gleeson-White, *Six Capitals: Or, Can Accountants Save the Planet? Rethinking Capitalism for the Twenty-First Century* (New York: W. W. Norton, 2015).

37. According to consultancy TruCost, "The total value of natural capital to society globally has been estimated to be up to $72 trillion per year according to the U.N. Environment Program. . . . If companies had to internalize all of the natural capital costs associated with their business, for example, as a result of increased regulations or new carbon taxes, their profit would be significantly at risk. The natural capital cost generated by the largest 1,200 companies in the world is more than twice their net income." GreenBiz Group and TruCost ESG, "State of Green Business 2019," 59–60, https://www.greenbiz.com/report/2019 -state-green-business-report, accessed April 10, 2020.

38. Herman E. Daly, *From Uneconomic Growth to a Steady-State Economy* (Cheltenham, UK and Northampton, MA: Edward Elgar, 2014); Daly and Farley, *Ecological Economics,* chap 2.

39. The data-analytics company INRIX keeps a good overview of congestion and its tragic economic and well-being costs, see http://inrix.com/scorecard/.

40. Stuchtey, Enkvist, and Zumwinkel, *A Good Disruption.*

41. Source for figure 4.3: P. Ekins, N. Hughes et al., *Resource Efficiency: Potential and Economic Implications, a Report of the International Resource Panel* (UNEP, 2017); IRP et al., *Global Resources Outlook 2019.*

42. Parrique et al., "Decoupling Debunked: Evidence and Arguments against Green Growth"; Jason Hickel and Giorgos Kallis, "Is Green Growth Possible?," *New Political Economy* (April 17, 2019): 1–18.

43. Mathis Wackernagel and Bert Beyers, *Ecological Footprint: Managing Our Biocapacity Budget* (Gabriola Island, BC: New Society Publishers, 2019), 69.

44. For a discussion of the ecological footprint methodology's weaknesses and strengths, see Alessandro Galli et al., "Questioning the Ecological Footprint," *Ecological Indicators* 69 (October 2016): 224–232; Thomas Wiedmann and John Barrett, "A Review of the Ecological Footprint Indicator—Perceptions and Methods," *Sustainability* 2, no. 6 (June 7, 2010): 1645–1693.

45. Global Footprint Network, "National Footprint Accounts," 30.

46. Source: Global Footprint Network, https://www.footprintnetwork.org/our-work/climate -change/, accessed October 1, 2019.

CHAPTER 5

1. Jorgen Randers, "Greenhouse Gas Emissions per Unit of Value Added ('GEVA')—A Corporate Guide to Voluntary Climate Action," *Energy Policy* 48 (September 2012): 46–55; Axel Haller, "Value Creation: A Core Concept of Integrated Reporting," in *Integrated*

Reporting (London: Springer, Palgrave Macmillan, 2016), 37–57; Axel Haller, Chris J. van Staden, and Cristina Landis, "Value Added as Part of Sustainability Reporting: Reporting on Distributional Fairness or Obfuscation?," *Journal of Business Ethics* 152, no. 3 (October 2018): 763–781.

2. See Grubler et al., "A Low Energy Demand Scenario for Meeting the 1.5°C Target," fig. 2; UNEP, *Emissions Gap Report*, fig. 5.2.

3. There are, of course, several barriers to recycling EEE materials at scale. See Christine Cole et al., "Assessing Barriers to Reuse of Electrical and Electronic Equipment, a UK Perspective," *Resources, Conservation & Recycling: X* 1 (June 2019): 1–10.

4. A footprint family is discussed in Alessandro Galli et al., "Integrating Ecological, Carbon and Water Footprint into a 'Footprint Family,'" *Ecological Indicators* 16 (May 2012): 100–112. At aggregate level the correlation between material throughput and ecological impacts is in the 0.70–0.88 range. Ester Voet, Lauran Oers, and Igor Nikolic, "Dematerialization: Not Just a Matter of Weight," *Journal of Industrial Ecology* 8, no. 4 (September 2004): 121–137; Moana Simas et al., "Correlation between Production and Consumption-Based Environmental Indicators," *Ecological Indicators* 76 (May 2017): 317–323.

5. Previous publications that have explored related visualizations include Jarmo Vehmas et al., *Europe in the Global Battle of Sustainability: Rebound Strikes Back? Advanced Sustainability Analysis* (Turku: Turun Kauppakorkeakoulu, 2003); Petri Tapio, "Towards a Theory of Decoupling: Degrees of Decoupling in the EU and the Case of Road Traffic in Finland between 1970 and 2001," *Transport Policy* 12, no. 2 (March 2005): 137–151; Mariana Conte Grand, "Carbon Emission Targets and Decoupling Indicators," *Ecological Indicators* 67 (August 2016): 649–56; van den Bergh, "Agrowth Instead of Anti- and Pro-Growth: Less Polarization, More Support for Sustainability/Climate Policies"; OECD, *Green Growth Indicators 2017*, OECD Green Growth Studies (OECD Publishing, 2017); Victor, *Managing without Growth*, fig 8.4.

6. OECD, *Green Growth Indicators 2017*.

7. Nathan W. Chan and Kenneth Gillingham, "The Microeconomic Theory of the Rebound Effect and Its Welfare Implications," *Journal of the Association of Environmental and Resource Economists* 2, no. 1 (March 2015): 133–159.

8. Harry D. Saunders, "The Khazzoom-Brookes Postulate and Neoclassical Growth," *Energy Journal* 13, no. 4 (1992); Harry D. Saunders, "Recent Evidence for Large Rebound," *Energy Journal* 36, no. 1 (2015).

9. David Owen, "The Efficiency Dilemma," *New Yorker* 20, no. 27 (2010): 78–80; Timan Santarius, "Green Growth Unravelled: How Rebound Effects Baffle Sustainability Targets When the Economy Keeps Growing," 2012, https://www.greengrowthknowledge.org/resource/green-growth-unravelled-how-rebound-effects-baffle-sustainability-targets-when-economy; Santarius, *Rethinking Climate and Energy Policies*; T. Parrique et al., "Decoupling Debunked: Evidence and Arguments against Green Growth" (European Environmental Bureau, 2019).

10. Lorna A. Greening, David L. Greene, and Carmen Difiglio, "Energy Efficiency and Consumption—the Rebound Effect—a Survey," *Energy Policy* 28, nos. 6–7 (2000): 389–401; Lee Schipper and Michael Grubb, "On the Rebound? Feedback between Energy Intensities and Energy Uses in IEA Countries," *Energy Policy* 28, no. 6–7 (2000): 367–388; Kenneth Gillingham et al., "The Rebound Effect Is Overplayed," *Nature* 493, no. 7433 (January 2013): 475–476; Inês M. L. Azevedo, "Consumer End-Use Energy Efficiency and Rebound Effects," *Annual Review of Environment and Resources* 39 (2014): 393–418.

11. Kenneth Gillingham, David Rapson, and Gernot Wagner, "The Rebound Effect and Energy Efficiency Policy," *Review of Environmental Economics and Policy* 10, no. 1 (2016): 68–88; Severin Borenstein, "A Microeconomic Framework for Evaluating Energy Efficiency Rebound and Some Implications," *Energy Journal* 36, no. 1 (2015).

12. Borenstein, "A Microeconomic Framework for Evaluating Energy Efficiency Rebound and Some Implications"; Amory B. Lovins et al., "Recalibrating Climate Prospects," *Environmental Research Letters* 14, no. 12 (December 2, 2019): 120201.

13. Gillingham et al., "The Rebound Effect Is Overplayed"; Weizsäcker, *Factor Five*, ch. 8; Lovins et al., "Recalibrating Climate Prospects."

14. See Per Espen Stoknes and Johan Rockström, "Redefining Green Growth within Planetary Boundaries," *Energy Research & Social Science* 44 (October 2018): 41–49; Corinne Le Quéré et al., "Drivers of Declining CO_2 Emissions in 18 Developed Economies," *Nature Climate Change* 9, no. 3 (March 2019): 213–17.

15. World Bank data: https://data.worldbank.org/country/japan, accessed April 10, 2020.

16. Good reviews of flaws and weaknesses with GDP as a measure of progress, are given in Joseph Stiglitz, Amartya Sen, and Jean-Paul Fitoussi, *Report of the Commission on the Measurement of Economic Performance and Social Progress* (Paris, 2009); Joseph E. Stiglitz, Jean-Paul Fitoussi, and Martine Durand, *Beyond GDP: Measuring What Counts for Economic and Social Performance* (OECD Publishing, Paris, 2018), https://www.oecd-ilibrary.org/content/publication/9789264307292-en; van den Bergh, "Agrowth Instead of Anti- and Pro-Growth."

17. See Stoknes, *Money & Soul*, chs. 5 and 11.

18. Gleeson-White, *Six Capitals*; and Diane Coyle, *GDP—A Brief but Affectionate History* (Princeton, NJ: Princeton University Press, 2015).

19. Stiglitz, Fitoussi, and Durand, *Beyond GDP*.

20. Thomas L. Friedman, "The Power of Green," New York Times, April 15, 2007; OECD, *Towards Green Growth* (OECD Green Growth Studies, OECD Publishing, Paris, 2011); UNEP, *Towards a Green Economy: Pathways to Sustainable Development and Poverty Eradication—A Synthesis for Policy Makers* (2011), www.unep.org/greeneconomy; Lisa Friedman, "What Is the Green New Deal? A Climate Proposal, Explained," *New York Times*, February 21, 2019.

21. OECD, *Towards Green Growth*.

22. Stoknes and Rockström, "Redefining Green Growth within Planetary Boundaries"; Maxton, Randers, and Suzuki, *Reinventing Prosperity*, 145.

23. There are about 12.2 billion global hectares (gha) of biologically productive area on Earth. Mathis Wackernagel et al., "National Natural Capital Accounting with the Ecological Footprint Concept," *Ecological Economics* 29, no. 3 (June 1999): 375–90; Wackernagel and Beyers, *Ecological Footprint*.

24. Planetary boundaries are presented in chapter 7, building on Johan Rockström et al., "A Safe Operating Space for Humanity," *Nature* 461, no. 7263 (September 24, 2009): 472–75; Will Steffen et al., "Planetary Boundaries: Guiding Human Development on a Changing Planet," *Science* 347, no. 6223 (2015): 1259855.

25. Thomas Piketty, "About Capital in the Twenty-First Century," *American Economic Review* 105, no. 5 (May 2015): 48–53.

26. We could add that g must be greater than 0 for there to be green growth: $rp > g > 0$. We get green degrowth if $rp > g$ and $g < 0$.

27. Stuchtey, Enkvist, and Zumwinkel, *A Good Disruption,* section 2.2; Ayres and Voudouris, "The Economic Growth Enigma."

28. For more on labor productivity and resource productivity, see Aaron E. Cobet and Gregory A. Wilson, "Comparing 50 Years of Labor Productivity in US and Foreign Manufacturing," *Monthly Lab. Rev.* 125 (2002): 51; IRP et al., "Global Resources Outlook 2019," 53; A. Stocker et al., *The Interaction of Resource and Labour Productivity: The Scientific Background* (Vienna: Sustainable Europe Research Institute, 2015), 14; Robert J. Gordon, "Is US Economic Growth Over? Faltering Innovation Confronts the Six Headwinds" (National Bureau of Economic Research, 2012); Gordon, *The Rise and Fall of American Growth.*

29. Herman E. Daly, "Economics in a Full World," *Scientific American* 293, no. 3 (2005): 100–107; Daly and Farley, *Ecological Economics*, 161.

30. Source for figure 5.2: by author. Data sources: Historic global GDP (USD 2011 PPP) has been reconstructed since 1900 by Maddison project database, https://www.rug.nl/ggdc/historicaldevelopment/maddison/releases/maddison-project-database-2018; ecological footprint accounts from Global Footprint Network, https://www.footprintnetwork.org/; and historic GHG emissions from IPCC https://www.ipcc.ch/report/ar5/wg3/, all accessed April 10, 2020. Author projections of GDP and green growth footprint are made near to the SSP1–SSP2 scenarios in the Shared Socioeconomic Pathways descriptions, building on the LED/P1 and S2/P3 scenarios in IPCC, *The Special Report on Global Warming of 1.5 °C (SR15)*, 2018.

31. Global Footprint Network, "National Footprint Accounts."

32. See Ernst von Weizsäcker, *Factor Five: Transforming the Global Economy through 80% Improvements in Resource Productivity* (London: Earthscan, 2009). More examples and references can be found in the Further Reading section.

33. "Changes in CO_2 emissions in the decade 2005–2015 differ from historical trends in that they are supported by the largest decreases in the fossil fuel share observed since 1960, and by the only decrease in energy use sustained over a decade." Quoted in Le Quéré et al., "Drivers of Declining CO2 Emissions in 18 Developed Economies." See also Stoknes and Rockström, "Redefining Green Growth within Planetary Boundaries"; OECD, *Green Growth Indicators 2017*; Jiandong Chen et al., "Decomposition and Decoupling Analysis of CO_2 Emissions in OECD," *Applied Energy* 231 (December 2018): 937–50; Richard Wood et al., "Beyond Peak Emission Transfers," *Climate Policy* (June 11, 2019): 1–14.

34. McAfee, *More from Less.*

35. Mark Carney, "Resolving the Climate Paradox" (Arthur Burns Memorial Lecture, Berlin, September 22, 2016); Task Force on Climate-Related Financial Disclosures, *Task Force on Climate-Related Financial Disclosures: Status Report 2019* (2019).

36. The healthy growth framework builds on John Elkington's classic triple bottom line (1999) concept but expands it into a flow component and a stock component. It also builds on Jonathan Porritt's five capitals model from the *Forum for the Future*, and on Integrated Reporting's framework of "Six Capitals" from Gleeson-White, *Six Capitals*. But it crystalizes their complex five and six capitals models into three strong, nonsubstitutable capital categories, and highlights only the most critical factors within each (labor, resource, and social productivity, as explained in chapter 7) to make it easier for quantification, comparison, and scalable navigation, all based on value-added. It also builds on the World Bank's *Wealth of Nations* and the UNEP's *Inclusive Wealth* approach, both applying the genuine savings methods (which works reasonably well for the productive/market economy in the left-most column). But the healthy growth approach does not apply shadow prices (or zero price) for core types of natural and social capital as these do. The novelty of the healthy growth approach is to underline the nonconvertible nature of these capitals to money values, and to instead use integrated metrics based on value-added. Rather than measuring all these capitals in dollars (like WB and UNEP which make them seem perfectly substitutable), I suggest that productive capital is measured in real $ per person, natural capital as the rate of change in the biocapacity reserve in gha, and social capital as rate of change in trust in government and society. A healthy economy requires that all three capital stocks have a non-negative rate of change over time. This can be achieved with healthy growth, where $rp > 5$ percent and $sp > 5$ percent. The main benefit is that these metrics scale up and down from companies/cities via countries to the global level, and sets the minimum sufficient rates of change in productivity to transform in a healthy direction (inside planetary boundaries) in time (by 2050 at the latest). See John Elkington, *Cannibals with Forks: The Triple Bottom Line of 21st Century Business*, reprint (Oxford: Capstone, 1999); Jonathon Porritt, *Capitalism as If the World Matters*, rev. ed (London: Earthscan, 2007); Gleeson-White, *Six Capitals*; Glenn-Marie Lange, Quentin Wodon, and Kevin Carey, *The Changing Wealth of Nations 2018: Building a Sustainable Future, World Bank* (Washington, DC: The World Bank, 2018); Shunsuke Managi and Pushpam Kumar, *Inclusive Wealth Report 2018: Measuring Progress towards Sustainability* (New York: Routledge, 2018).

37. The *Inclusive Wealth Report* uses so-called "shadow prices" to calculate changes in human and natural capitals. That means that the prices are derived from research and guesstimates rather than observed market prices. Also, it means that countries' natural capital is accounted for in $ (and easily dominated by oil-price fluctuations) while social capital is not priced at all. This contrasts with my healthy growth approach, where we'll use integrated metrics (sufficient rates of change in resource productivity value-added $/ton and in social productivity value-added $ adjusted for inequality) rather than shadow prices. For global results on the 1992–2014 period, see Managi and Kumar, *Inclusive Wealth Report 2018*, 15.

38. The New Climate Economy Report (NCER) study of 2018 finds a direct $26 trillion economic gain through 2030 compared with business as usual as a conservative estimate; the LUT University and Energy Watch Group (EWG) by Ram et al. report that a global 100 percent renewable energy system can be achieved with zero GHG emissions before 2050 and more cost-effectively than the current fossil fuel system, a conclusion supported by DNV GL in *Energy Transition Outlook 2019*; and the Business & Sustainable Development Commission (BSDC) report finds $12–28 trn by 2030 in business savings and revenue. See NCER, *Unlocking the Inclusive Growth Story of the 21st Century*; Ram et al., *Global Energy System Based on 100% Renewable Energy*; BSDC, "Better Business Better World" (London: Systemiq, January 2017).

39. OECD, *In It Together: Why Less Inequality Benefits All* (OECD, 2015); IMF, *Fostering Inclusive Growth: G-20 Leaders' Summit* (IMF, June 26, 2017); Jonathan David Ostry, Andrew Berg, and Charalambos G. Tsangarides, *Redistribution, Inequality, and Growth* (IMF Staff Discussion Note, February 2014).

CHAPTER 6

1. For info on the Plastic Bank project in Indonesia, see "Plastic Bank—the $4 Trillion Economy That Will Solve Indonesia's Plastic Crisis," *WowShack*, November 16, 2018, https://www.wowshack.com/plastic-bank-the-4-trillion-economy-that-will-solve-indonesias-plastic-crisis/.

2. Steven Pinker, *Enlightenment Now: The Case for Reason, Science, Humanism, and Progress* (New York: Viking, 2018); Hans Rosling, Ola Rosling, and Anna Rönnlund, *Factfulness: Ten Reasons We're Wrong about the World* (New York: Flatiron, 2018).

3. The seven-region concept is based on clustering countries with similar characteristics and is used in the global systems model Earth3 for a 2018 report by *Limits to Growth* coauthor Jorgen Randers, planetary boundary originator Johan Rockström, me, and others. The report analyzed whether achieving the SDGs within planetary boundaries is possible. See Randers et al., "*Transformation Is Feasible—How to Achieve the 17 SDGs within Planetary Boundaries.*"

4. $40/day is the same as (40*365) = $15,000 GDP per person per year in 2011$ PPP. All $ numbers in this chapter and figures refer to 2011 values of US dollars at purchasing power parity (PPP). Other currencies are converted to USD at the dollar's value in 2011,

according to the purchasing value of the local currency, *not* the running market exchange rate. This means that the USD reflect the same purchasing power equivalent in the United States of what the other country's currency would have had in that country. Purchasing power exchange rates are such that it would cost exactly the same number of US dollars to buy rupees and then buy a basket of goods in the Indian market as it would cost to purchase the same goods directly with dollars in the US. Thus, PPP rates tend to be higher for poor country currencies than the market exchange rate.

5. There is of course no guarantee of "business as usual" being able to go on to 2050 and beyond, and—as we will see in the last chapter—this conventional scenario may also undermine the sustainability of the very good news it otherwise represents, by pushing life-supporting Earth systems into irreversible decline well before 2050. But for now let's recognize this good news in itself.

6. For more detail on SDG performance, see Randers et al., *Transformation Is Feasible—How to Achieve the 17 SDGs within Planetary Boundaries*. For additional indicators and a full national SDG Dashboard, see Jeffrey Sachs et al., *SDG Index & Dashboards: A Global Report* (Bertelsmann Stiftung, 2016).

7. See David Collste et al., "The Empirical Bases for the Earth3 Model: Technical Notes on the Sustainable Development Goals and Planetary Boundaries," EarthArXiv, October 1, 2018.

8. See, for instance, Guy Standing, *The Precariat: The New Dangerous Class* (London: Bloomsbury Academic, 2011); Ostry, Berg, and Tsangarides, "Redistribution, Inequality, and Growth"; Joseph E. Stiglitz, *The Price of Inequality* (New York: Norton & Company, 2013); OECD, *In It Together*; Kate E. Pickett and Richard G. Wilkinson, "Income Inequality and Health: A Causal Review," *Social Science & Medicine* 128 (March 2015): 316–26; Piketty, "About Capital in the Twenty-First Century"; IMF, "Fostering Inclusive Growth: G-20 Leaders' Summit."

9. Figure 6.1 from Alvaredo et al., *World Inequality Report 2018*, 51, figure 2.1.4.

10. In the 1984–2013 period, the compound average growth in US median household income was only 0.3 percent per year. Yet in the period from 2013 to 2017, median income growth rate was 3.3 percent per year, breaking the "flat" trend. Hence, over the 1984–2017 period there has been only 0.6 percent average growth, much lower than mean incomes and GDP per person. See US Census Bureau, "Real Median Household Income in the United States [MEHOINUSA672N]," retrieved from FRED, Federal Reserve Bank of St. Louis, https://fred.stlouisfed.org/series/MEHOINUSA672N, accessed August 10, 2019. See also Brian Nolan, "Stagnating Median Incomes despite Economic Growth: Explaining the Divergence in 27 OECD Countries," *Economics* 16 (2019): 17.

11. See Susan Milligan, "Stretched Thin: The Continuing Government Shutdown Highlights the Fact That Most Americans Have Little to No Savings," *US News & World Report*, January 11, 2019.

12. Stiglitz, *The Price of Inequality*; Ostry, Berg, and Tsangarides, "Redistribution, Inequality, and Growth."

13. Thomas Piketty, *Capital in the Twenty-First Century* (Cambridge, MA: The Belknap Press of Harvard University Press, 2014).

14. R. Samans et al., *The Inclusive Development Index 2018: Summary and Data Highlights* (Geneva, Switzerland: World Economic Forum, 2019).

15. Alex Cobham and Andy Sumner, "Is It All about the Tails? The Palma Measure of Income Inequality," *SSRN Electronic Journal* (January 2013); Alex Cobham and Andy Sumner, "Putting the Gini Back in the Bottle? 'The Palma' as a Policy-Relevant Measure of Inequality" (London: King's College London, March 15, 2013); Ija Trapeznikova, "Measuring Income Inequality," *IZA World of Labor* (July 2019).

16. José Gabriel Palma, "Globalizing Inequality: 'Centrifugal' and 'Centripetal' Forces at Work" (DESA working paper 35, United Nations, 2006); José Gabriel Palma, "Homogeneous Middles vs. Heterogeneous Tails, and the End of the 'Inverted-U': It's All about the Share of the Rich," *Development and Change* 42, no. 1 (2011): 87–153.

17. Erik Bengtsson and Daniel Waldenström, "Capital Shares and Income Inequality: Evidence from the Long Run," *Journal of Economic History* 78, no. 3 (2018): 712–743.

18. Alvaredo et al., *World Inequality Report 2018*, figure E8.

19. Casey Quackenbush, "The World's Top 26 Billionaires Now Own as Much as the Poorest 3.8 Billion, Says Oxfam," *Time*, January 21, 2019, https://time.com/5508393/global-wealth-inequality-widens-oxfam/.

20. According to a study by Careerbuilder: http://press.careerbuilder.com/2017-08-24-Living-Paycheck-to-Paycheck-is-a-Way-of-Life-for-Majority-of-U-S-Workers-According-to-New-CareerBuilder-Survey.

21. Standing, *The Precariat*.

22. Quote from https://www.unitedforalice.org/home. All reports are available from https://www.unitedforalice.org/all-reports.

23. Guy Standing, *A Precariat Charter: From Denizens to Citizens* (London and New York: Bloomsbury, 2014); Stiglitz, *The Price of Inequality*.

24. Arthur M. Okun, *Equality and Efficiency: The Big Tradeoff*, rev. ed. (1975; Washington, DC: Brookings Institution Press, 2015).

25. Joseph E. Stiglitz, *Rewriting the Rules of the American Economy: An Agenda for Growth and Shared Prosperity* (New York: W. W. Norton, 2015).

26. See Ostry, Berg, and Tsangarides, "Redistribution, Inequality, and Growth"; Prakash Loungani, "The Power of Two: Inclusive Growth and the IMF," *Intereconomics* 52, no. 2 (March 2017): 92–99; IMF, "Fostering Inclusive Growth: G-20 Leaders' Summit"; OECD, *Growing Unequal? Income Distribution and Poverty in OECD Countries* (Paris: OECD, 2008); OECD, *In It Together*; M. Forster, W. Chen, and A. Llenanozal, *Divided We Stand: Why*

Inequality Keeps Rising (Paris: OECD, 2011); World Bank, "Inclusive Green Growth The Pathway to Sustainable Development"; World Bank, *Inclusion Matters: The Foundation for Shared Prosperity* (Washington, DC: The World Bank, 2013).

27. See discussion in OECD, *In It Together*, section 2.4.

28. IMF, "Fostering Inclusive Growth: G-20 Leaders' Summit," 9–10; OECD, *In It Together*, section 1.4; Standing, *The Precariat*.

29. IMF, "Fostering Inclusive Growth: G-20 Leaders' Summit," 20; OECD, *In It Together*, 81, note 1. See also B. Keith Payne, Jazmin L. Brown-Iannuzzi, and Jason W. Hannay, "Economic Inequality Increases Risk Taking," *Proceedings of the National Academy of Sciences* 114, no. 18 (May 2, 2017): 4643–4648.

30. Alberto Alesina and Dani Rodrik, "Distributive Politics and Economic Growth," *Quarterly Journal of Economics* 109, no. 2 (1994): 465–490; Simplice Asongu and Rangan Gupta, "Trust and Quality of Growth: A Note," *Economics Bulletin* 36, no. 3 (2016): 1854–1867; IMF, "Fostering Inclusive Growth: G-20 Leaders' Summit," 10.

31. Hector Rufrancos and Madeleine Power, "Income Inequality and Crime: A Review and Explanation of the Timeseries Evidence," *Sociology and Criminology-Open Access* 1, no. 1 (2013).

32. Jonas Minet Kinge et al., "Association of Household Income With Life Expectancy and Cause-Specific Mortality in Norway, 2005–2015," *JAMA* (May 13, 2019).

33. Jonathan K. Burns, Andrew Tomita, and Amy S. Kapadia, "Income Inequality and Schizophrenia: Increased Schizophrenia Incidence in Countries with High Levels of Income Inequality," *International Journal of Social Psychiatry* 60, no. 2 (2014): 185–196; Sheri L. Johnson, Erik Wibbels, and Richard Wilkinson, "Economic Inequality Is Related to Cross-National Prevalence of Psychotic Symptoms," *Social Psychiatry and Psychiatric Epidemiology* 50, no. 12 (2015): 1799–1807; Erick Messias, William W. Eaton, and Amy N. Grooms, "Economic Grand Rounds: Income Inequality and Depression Prevalence across the United States: An Ecological Study," *Psychiatric Services* 62, no. 7 (2011): 710–712.

34. Source for figure 6.2: Richard G. Wilkinson and Kate E. Pickett, "Income Inequality and Social Dysfunction," *Annual Review of Sociology* 35, no. 1 (August 2009): 493–511.

35. Pickett and Wilkinson, "Income Inequality and Health"; Richard G. Wilkinson and Kate E. Pickett, "The Enemy between Us: The Psychological and Social Costs of Inequality," *European Journal of Social Psychology* 47, no. 1 (2017): 11–24; Wilkinson and Pickett, "Income Inequality and Social Dysfunction."

36. Lange, Wodon, and Carey, *The Changing Wealth of Nations 2018*, 14; Managi and Kumar, *Inclusive Wealth Report 2018*, 35.

37. OECD, *In It Together*; Ostry, Berg, and Tsangarides, "Redistribution, Inequality, and Growth"; Peter Hoeller, Isabelle Joumard, and Isabell Koske, "Reducing Income Inequality While Boosting Economic Growth: Can It Be Done? Evidence from OECD Countries," *Singapore Economic Review* 59, no. 01 (2014); M. Mlachila, R. Tapsoba, and S. Tapsoba, *A Quality of Growth Index for Developing Countries: A Proposal* (Washington, DC:

International Monetary Fund, 2014); Christoph Boehringer, Xaquin Garcia-Muros, and Mikel González-Eguino, "Greener and Fairer: A Progressive Environmental Tax Reform for Spain" (Oldenburg Discussion Papers in Economics, 2019); Stiglitz, *The Price of Inequality*.

38. OECD, *In It Together*, ch. 2.

39. A too big topic to cover in this chapter, but it is an obvious fact that the power of corporate money and private wealth shifts the balance of power in policy-making away from the ideal of majority representative democracy. See for instance Martin Gilens and Benjamin Page, "Testing Theories of American Politics: Elites, Interest Groups, and Average Citizens," *Perspectives on Politics* (Cambridge: Cambridge University Press, 2014); Stiglitz, *The Price of Inequality*; Stiglitz, *Rewriting the Rules of the American Economy: An Agenda for Growth and Shared Prosperity*.

40. See "Table 3: Inequality-Adjusted Human Development Index," http://hdr.undp.org/en/composite/IHDI, accessed February 1, 2019.

41. Regarding the US's post-tax Palma ratio of 3.0 in 2014: data source downloaded from World Inequality Database (www.wid.world) on April 24, 2019, using variable "adi-inc_992_j_US" and formula: Palma ratio = p90p100 / (p0p40 × 4). This is significantly higher than, for instance, the OECD estimate on Palma ratio of 1.8 from OECD Income Distribution Database (IDD), http://stats.oecd.org/Index.aspx?DataSetCode=IDD, accessed April 24, 2019. This latter is based on household survey data that have several limitations on accuracy, rather than tax records as the WID data uses. See discussion of methodology in Alvaredo et al., *World Inequality Report 2018*, 27–33.

42. HDI-inequality adjusted, http://www.hdr.undp.org/en/composite/IHDI, accessed April 10, 2020; OECD's Quality of life/satisfaction index, http://www.oecdbetterlifeindex.org/, accessed April 10, 2020. In happiness: John F. Helliwell, Richard Layard, and Jeffrey Sachs, *World Happiness Report 2019* (New York: Sustainable Development Solutions Network, 2019).

43. Karl Ove Moene, "Scandinavian Equality: A Prime Example of Protection without Protectionism," in *The Quest for Security: Protection Without Protectionism and the Challenge of Global Governance*, ed. Joseph E Stiglitz and Mary Kaldor (New York: Columbia University Press, 2013), 48–74; Erling Barth and Karl Ove Moene, "The Equality Multiplier: How Wage Compression and Welfare Empowerment Interact," *Journal of the European Economic Association* 14, no. 5 (2016): 1011–1037.

44. Torben M. Andersen et al., "The Nordic Model. Embracing Globalization and Sharing Risks," *ETLA B* 232 (2007); John Magne Skjelvik, Karin Ibenholt, and Annegrete Bruvoll, *Greening the Economy*, TemaNord (Nordic Council of Ministers, 2011), http://urn.kb.se/resolve?urn=urn:nbn:se:norden:org:diva-1824.

45. For a nuanced discussion, see Torben M. Andersen and Jonas Maibom, "The Big Trade-off between Efficiency and Equity—Is It There?," CEPR Discussion Paper No. DP11189 (March 2016).

46. Ronald Reagan's across-the-board supply-side tax cuts became fully effective at the start of 1983. Data source: Post-tax national income, Total population, average income, adults, equal split, Dollar $ PPP constant (2017) USA, variable name: "adiinc_992_j_US." Downloaded from www.wid.world on April 24, 2019.

47. Euromonitor International, Lydia Gordon, and Fransua Razvadauskas, "Income Inequality Ranking of the World's Major Cities" (Euromonitor International, 2017), https://blog .euromonitor.com/income-inequality-ranking-worlds-major-cities/, accessed July 1, 2019.

48. To do so, the vertical axis that is used for countries to show the rate of change in the GDP should be replaced with rate of change in gross value added (GVA). GVA is the measure of the value of goods and services produced in an area, industry, or sector of an economy.

49. Labor share data from 1980 to 2015 in Marta Guerriero, "The Labor Share of Income around the World: Evidence from a Panel Dataset," ADBI Working Paper 920 (February 1, 2019); Eoin Flaherty and Seán Ó Riain, "Labour's Declining Share of National Income in Ireland and Denmark," *Socio-Economic Review* 17, no. 2 (2019); ILO and OECD, *The Labour Share in G20 Economies* (Report Prepared for the G20 Employment Working Group, Antalya, Turkey, February, 2015).

50. Haller, van Staden, and Landis, "Value Added as Part of Sustainability Reporting."

51. See *The Oxford Dictionary of Economics*, 5 ed. (Oxford: Oxford University Press, 2017): "Value added."

52. See value distributed chart, p. 22: https://www.tomra.com/en/investor-relations/ disclosures/tomra-annual-report-2013_31032014, accessed January 1, 2018.

53. Haller, "Value Creation: A Core Concept of Integrated Reporting"; Haller, van Staden, and Landis, "Value Added as Part of Sustainability Reporting."

54. Managi and Kumar, *Inclusive Wealth Report 2018*; Lange, Wodon, and Carey, *The Changing Wealth of Nations 2018*.

CHAPTER 7

1. The International Integrated Reporting Framework (IIRC 2013) describes integrated metrics as: "quantitative indicators that help increase comparability and are particularly helpful in expressing and reporting against targets. Common characteristics . . . include that they are connected (e.g. they display connectivity between financial and other information) . . . and presented for multiple periods (e.g. three or more periods) to provide an appreciation of trends." IIRC, *The International <IR> Framework* (Integrated Reporting Committee, 2013). That is why I support *rates of change* in resource productivity and social productivity over time, because they connect and integrate the key business interest of creating value-added/gross profits, with measurable critical environmental and social impacts.

2. This approach builds on a shift in moral philosophy, from the standard utilitarian ethics of conventional economics to a Kantian (deontological) ethics. See Mads Greaker et al., "A

Kantian Approach to Sustainable Development Indicators for Climate Change," *Ecological Economics* 91 (July 2013): 10–18.

3. Source for figure 7.2: figure by Lokrantz/Azote AB, modified by author. Scientific sources from: Rockström et al., "A Safe Operating Space for Humanity"; Steffen et al., "Planetary Boundaries"; Steffen et al., "The Trajectory of the Anthropocene"; Will Steffen et al., "Trajectories of the Earth System in the Anthropocene," *Proceedings of the National Academy of Sciences* 115, no. 33 (August 14, 2018): 8252–8259.

4. Companies can now get certified science-based targets from the Science-Based Targets Initiative (SBTI). For more info, see CDP, UN Global Compact, WRI and WWF, "Science-Based Targets," http://sciencebasedtargets.org/, accessed April 10, 2020; Randers, "Greenhouse Gas Emissions per Unit of Value Added ('GEVA')—A Corporate Guide to Voluntary Climate Action"; Oskar Krabbe et al., "Aligning Corporate Greenhouse-Gas Emissions Targets with Climate Goals," *Nature Climate Change* 5, no. 12 (2015): 1057–60; Steffen et al., "Planetary Boundaries."

5. Carbon productivity as a headline indicator: OECD, *Green Growth Indicators 2017*; Global Commission on the Economy and Climate, *Seizing the Global Opportunity: The New Climate Economy Report* (Washington, DC: The Global Commission on the Economy and Climate, 2015). Carbon is ~60 percent of humanity's footprint, see Wackernagel and Beyers, *Ecological Footprint*, 34.

6. Simas et al., "Correlation between Production and Consumption-Based Environmental Indicators."

7. Hans Joachim Schellnhuber et al., *Turn down the Heat—Why a 4 C Warmer World Must Be Avoided* (Washington, DC: World Bank, 2012).

8. The standards for greenhouse gas accounting (Scopes 1: direct; 2: energy; and 3: up and down the value chain) are described by the GHG Protocol: see http://ghgprotocol.org, accessed April 10, 2020.

9. Data source: M. Crippa et al., "Fossil CO_2 and GHG Emissions of All World Countries, 2019 EDGAR," Publications Office of the European Union, 2019. The report and database use the World Bank's GDP numbers in 2011$ PPP, and 2015 is the last year available for country-level GHG emissions.

10. On the three approaches (including value added) to calculating GDP, and its history, see Coyle, *GDP*; Gleeson-White, *Six Capitals*. On the main and many shortcomings of GDP, a good overview is also given by van den Bergh, "Agrowth Instead of Anti-and Pro-Growth," table 1.

11. By real GDP, I refer to output measured in constant prices—in other words, adjusted for inflation. Also, when doing international comparisons, economists most often use purchasing power parity (PPP) US dollars. In this book I use (unless stated otherwise) 2011$ PPP, as compiled by the leading database Penn World Tables 9.1. Robert C. Feenstra, Robert Inklaar, and Marcel P Timmer, "The Next Generation of the Penn World Table," *American Economic Review* 105, no. 10 (2015): 3150–3182.

12. See "GDP Growth (Annual %)," World Bank, https://data.worldbank.org/indicator/ny.gdp.mktp.kd.zg, accessed April 10, 2020.

13. Gordon, *The Rise and Fall of American Growth*; Lant Pritchett and Lawrence Summers, *Asiaphoria Meets Regression to the Mean* (Cambridge, MA: National Bureau of Economic Research, October 2014); Jorgen Randers, "How Fast Will China Grow towards 2030?," *World Economics* (June 2016); Jorgen Randers, *2052: A Global Forecast for the Next Forty Years* (White River Junction, VT: Chelsea Green Publishing, 2012); DNV GL, "Energy Transition Outlook—A Global and Regional Forecast of the Energy Transition to 2050."

14. Intergovernmental Panel on Climate Change, *Global Warming of 1.5°C* (Intergovernmental Panel on Climate Change, 2018), 108, table 2.2.

15. The wide range in estimates of remaining carbon budgets and pathways is largely explained by different assumptions in socioeconomic pathways (SSPs), levels of probability of succeeding (i.e. 50 percent, 66 percent, or 90 percent likelihood), assumptions about how much carbon capture and storage (negative emissions) we will see during the second half of this century, and finally—on a country level—how much more rich countries should cut due to historical responsibility than the poor countries. See also Johan Rockström et al., "A Roadmap for Rapid Decarbonization," *Science* 355, no. 6331 (2017): 1269–1271.

16. Hawken, *Drawdown*.

17. Miklós Antal and Jeroen van den Bergh, "Green Growth and Climate Change: Conceptual and Empirical Considerations," *Climate Policy* 16, no. 2 (2014): 165–177; Elmar Kriegler et al., "What Does the 2 C Target Imply for a Global Climate Agreement in 2020? The LIMITS Study on Durban Platform Scenarios," *Climate Change Economics* 4, no. 04 (2013): 1340008; New Climate Economy Report, *Better Growth, Better Climate: The New Climate Economy Report: The Global Report* (New Climate Economy, 2014); Randers, "Greenhouse Gas Emissions per Unit of Value Added ('GEVA')—A Corporate Guide to Voluntary Climate Action"; UNEP, *Towards a Green Economy: Pathways to Sustainable Development and Poverty Eradication*; Deep Decarbonization Pathways Project, *Pathways to Deep Decarbonization 2015 Report* (SDSN-IDDRI, 2015).

18. Other authors have calculated that higher rates than 5 percent per year are needed. A study by Rockström et al. suggests a "Carbon Law" of 6–7 percent CO_2e reductions pa, which would mean 8–9 percent per year carbon productivity rate starting in 2020 to have a more than 50 percent chance of reaching a 1.5°C target (Rockström et al., "A Roadmap for Rapid Decarbonization"). Tim Jackson calculated an even higher rate for rich countries, at 9–11 percent per year (in Jackson, *Prosperity without Growth*) but later raised the bar to 14–20 percent per year. Tim Jackson and Peter A. Victor, "Unraveling the Claims for (and against) Green Growth," *Science* 366, no. 6468 (2019): 950–951.

19. In a future, more complete healthy growth accounting, I suggest using four components to detail the broad footprint denominator in *rp*: capro (rate of change, RoC, in VA/tCO$_{2e}$), nupro (RoC VA/tons of nutrient loading), biopro (RoC VA/gha biocapacity footprint),

fopro (RoC VA/ha old-growth forest loss). These correspond to the four planetary bound-aries in the red high-risk zone. In many areas, water productivity, or wapro (RoC VA/tons of freshwater) would be very important, but it is not a global boundary yet in the red. One way to combine them all into a single rp to measure overall green growth could be to calculate the geometric mean of whichever are the most material with reliable data on capro, nupro, fopro, biopro, wapro. These issues are outside the scope of this chapter, and will be addressed in forthcoming publications.

The one established indicator that currently comes the closest to all these taken together, however imperfectly, is probably the ecological footprint (see Wackernagel and Beyers, *Ecological Footprint,* loc. 1264). But not many actors (companies, cities, or countries) yet account for their ecological footprint, nutrient, forest, or water use. The most available data is currently on GHG emissions according to the CDP/GHG protocol standards. Hence, the simplest current way to calculate rp with one number is to use capro as a proxy for rp.

20. Genuine green growth (GGG) is different from simply green growth (where $\Delta GDP > 0 > \Delta EF$, i.e., an absolute decoupling). It is also distinct from green degrowth ($\Delta GDP < 0$ and $\Delta EF < 0$), as well as from gray growth ($\Delta GDP > 0$ and $\Delta EF > 0$, which only gives, at best, a relative decoupling).

21. Why does $rp = 5.3$ percent? Assume a value creation in year 1 of $100 and a footprint of 100 tons CO_2 emissions. The resource productivity in year 1 is then $RP_1 = \$100/100$ t$= 1$ $/t. That means in order to make $1 of value creation, 1 ton is emitted. Next year, the GDP is down 1 percent to $99 and emissions down 6 percent to 94t. $RP_2 = \$99/94t = 1.053$ $/t. This gives a change in resource productivity $\Delta RP = RP_2 - RP_1 = (1.053-1)\$/t = 0.053$ $/t. And a rate of change in $rp = \Delta RP/RP_1 = 0.053/1 = 5.3$ percent.

22. Kallis, *Degrowth*; Hickel and Kallis, "Is Green Growth Possible?"; Victor, *Managing without Growth*.

23. Using the same procedure as in the footnote above, the example means value creation growth from $100 to $107 in year 2. While footprint changes from 100t to 101t. $RP_1 = \$100/100$ t $= 1$ $/t. $RP_2 = \$107/101t = 1.0594$ $/t. $\Delta RP = (1.0594-1)$, which means a rate of change of 5.9 percent and can be rounded to 6 percent. If we use continuous compounding rather than discreet compounded annual growth rates (CAGR) we can use the natural logarithm and find that $\ln(1.0594/1)/1 = 5.8$ percent. But as value creation is usually determined annually, in this book I use the discreet CAGR approach for resource productivity.

24. India's average footprint per person (gha/p) is currently way below the global: In 2016 India had an EF per person $= 1.2$ gha/p, while the global average is 2.7 gha/p. The Earth's total biocapacity per year is currently (12 bn gha / 7.8 bn p =) 1.5 gha/p. That means humanity's current footprint is equal to (2.7 / 1.7=) 1.8 planet equivalents. See https://www.footprintnetwork.org/resources/glossary/, accessed April 10, 2020.

25. Randers et al., *Transformation Is Feasible—How to Achieve the 17 SDGs within Planetary Boundaries*, 38.

26. Source: http://www.ellafitzgerald.com/about/quotes, accessed April 10, 2020.

27. Rahul Anand, Saurabh Mishra, and Shanaka J. Peiris, *Inclusive Growth: Measurement and Determinants* (Washington, DC: International Monetary Fund, 2013); World Bank, "Inclusive Green Growth: The Pathway to Sustainable Development"; U. Narloch, T. Kozluk, and A. Lloyd, *Measuring Inclusive Green Growth at the Country Level. Taking Stock of Measurement Approaches and Indicators* (GGKP Working Paper 2, February 2016); IMF, "Fostering Inclusive Growth: G-20 Leaders' Summit."

28. Naoki Kondo et al., "Income Inequality and Health: The Role of Population Size, Inequality Threshold, Period Effects and Lag Effects," *Journal of Epidemiology and Community Health* 66, no. 6 (June 2012): e11; Pickett and Wilkinson, "Income Inequality and Health," 321.

29. UNDP source: http://hdr.undp.org/en/composite/IHDI, accessed April 10, 2020.

30. Using ranking from Helliwell, Layard, and Sachs, *World Happiness Report 2019*, and Palma ratios from UNDP, http://hdr.undp.org/en/composite/IHDI.

31. Source: UN SDG Knowledge Platform, https://sustainabledevelopment.un.org/topics/sustainabledevelopmentgoals, accessed April 10, 2020.

32. Alvaredo et al., *World Inequality Report 2018*, 40.

33. Michael W. Doyle and Joseph E. Stiglitz, "Eliminating Extreme Inequality: A Sustainable Development Goal, 2015–2030," *Ethics & International Affairs* 28, no. 1 (2014): 5–13.

34. Moene, "Scandinavian Equality: A Prime Example of Protection without Protectionism"; Barth and Moene, "The Equality Multiplier: How Wage Compression and Welfare Empowerment Interact."

35. Sachs et al., *SDG Index & Dashboards: A Global Report,* table 7. See also Oxfam, *Reward Work Not Wealth* (Oxford: Oxfam International, 2018).

36. Lars Engberg-Pedersen, *Development Goals Post 2015: Reduce Inequality* (DIIS: Copenhagen, 2013); Pickett and Wilkinson, "Income Inequality and Health"; Wilkinson and Pickett, "The Enemy between Us: The Psychological and Social Costs of Inequality"; José Gabriel Palma and Joseph E. Stiglitz, "Do Nations Just Get the Inequality They Deserve? The 'Palma Ratio' Re-Examined," in *Inequality and Growth: Patterns and Policy* (Berlin: Springer, 2016), 35–97.

37. Engberg-Pedersen, at the Danish Institute for International Studies, suggests halving the part of the Palma ratio that exceeds one in 2030 compared to 2010. Engberg-Pedersen, *Development Goals Post 2015: Reduce Inequality.*

38. For discussion of global GDP rates per world region and globally see DNV GL, "Energy Transition Outlook 2019," 35.

39. These numbers are calculated by the author, based on tax data sources as compiled in the World Inequality Database. See www.wid.world, variable adiinc_992_j_US for US,

p90p100 / p0p40, accessed December 29, 2019. There are other measurements of Palma that give lower values for the US and many other countries, often based on household survey data. But tax data for income inequality is seen as more correct and robust; see discussion in Alvaredo et al., *World Inequality Report 2018*, 28–31.

40. The definition of $rp > 5$ and $sp > 5$ rests on the robust assumption that global GDP declines from average of 3% to an average long-term growth rate of near 2 percent by 2050. See note 13 in this chapter.

41. Kevin Anderson and A. Bows-Larkin, "Avoiding Dangerous Climate Change Demands De-Growth Strategies from Wealthier Nations," *Kevinanderson.Info*, November 25, 2013, http://kevinanderson.info/blog/avoiding-dangerous-climate-change-demands-de-growth -strategies-from-wealthier-nations/; Jackson, *Prosperity without Growth*; Sylvia Lorek and Joachim H. Spangenberg, "Sustainable Consumption within a Sustainable Economy— beyond Green Growth and Green Economies," *Journal of Cleaner Production* 63 (January 2014): 33–44; Enno Schröder and Servaas Storm, *Economic Growth and Carbon Emissions: The Road to "Hothouse Earth" Is Paved with Good Intentions* (New York: Institute for New Economic Thinking, November 2018); Hickel and Kallis, "Is Green Growth Possible?"

42. Source for figure 7.5: by the author. Data source: OECD stats, *Green Growth Indicators*, Production-based CO_2 productivity, GDP in USD 2010 PPP/kgCO_2 emissions, https://stats .oecd.org/Index.aspx?DataSetCode=GREEN_GROWTH, accessed April 25, 2020.

43. These are rates of change (CAGR) from 2010 to latest available year (2017–2018) from OECD stats, *Green Growth Indicators*, https://stats.oecd.org/Index.aspx?DataSetCode =GREEN_GROWTH, Production-based CO_2 productivity, GDP in USD 2010 per unit of energy-related CO_2 emissions, accessed November 17, 2019.

44. EU target source: Sam Morgan, "EU Ministers Fudge 2030 Climate Target Lines," *Euractiv*, October 6, 2019, https://www.euractiv.com/section/climate-environment/news/ eu-ministers-fudge-2030-climate-target-lines/. Capro rates: author's calculations based on data from OECD stats.

45. Olle Björk et al., *Making the Environment Count* (TemaNord 2016:507), https://norden .diva-portal.org/smash/get/diva2:915431/FULLTEXT01.pdf, accessed April 10, 2020; Skjelvik, Ibenholt, and Bruvoll, *Greening the Economy*; Benjamin Sovacool, "Contestation, Contingency, and Justice in the Nordic Low-Carbon Energy Transition," *Energy Policy* 102 (March 2017): 569–582.

46. IEA, *Energy Policies of IEA Countries Sweden 2019 Review* (IEA, 2019), www.iea.org.

47. IEA, *Energy Policies of IEA Countries Denmark 2017 Review* (IEA, 2017), www.iea.org.

48. IEA, *Energy Policies of IEA Countries Norway 2017 Review*.

49. Frauke Urban and Johan Nordensvärd. "Low Carbon Energy Transitions in the Nordic Countries: Evidence from the Environmental Kuznets Curve," *Energies* 11, no. 9 (August 23, 2018): 1–17.

50. GDP and capro from OECD Stats Green Growth Indicators. Kristina Mohlin et al., "Factoring in the Forgotten Role of Renewables in CO2 Emission Trends Using Decomposition Analysis," *Energy Policy* 116 (2018): 290–296.

51. Fergus Green and Nicholas Stern, "China's Changing Economy: Implications for Its Carbon Dioxide Emissions," *Climate Policy*, March 16, 2016, 1–15; Shinwei Ng, Nick Mabey, and Jonathan Gaventa, "Pulling Ahead on Clean Technology; China's 13th Five Year Plan Challenges EU's Low Carbon Competitiveness," E3G.org, March 2016. Recently however, CO_2 emissions have been rising again (2017–2018). For a discussion, see: https://www .carbonbrief.org/guest-post-why-chinas-co2-emissions-grew-less-than-feared-in-2019.

52. At the time of writing, the latest year of data was 2014. Data source: Disposable adult incomes as given by World Inequality Database, www.wid.world, variable adiinc_992_j_ US for US, p90p100/p0p40, accessed December 29, 2019. Author's calculations of Palma ratio.

53. Source for table 7.1: 2014 USA tax data for 2014 from World Inequality Database and author scenarios to 2050, assuming low immigration, a stable population (to simplify) and 2.5 percent GDP growth, in constant 2017 US dollars.

54. Source for figure 7.7: author. Data source: World Inequality Database, with GDP USD 2011 PPP data from World Penn Tables, WPT9.1

55. There is, however, some recent survey data showing pickup of median wages in 2014–2016 (4.1 percent per year), but declining again in 2016–2018 (1.1 percent per year). See US Census Bureau, Real Median Household Income in the United States (MEHOINU-SA672N), https://fred.stlouisfed.org/series/MEHOINUSA672N, accessed December 29, 2019.

56. Source for GDP figures: Penn World Tables, version 9.1, using variable "rgdpna"; Feenstra et al., "The Next Generation of the Penn World Table."

57. The Office of National Statistics gives an overview of UK inequality since 1977 here: https://www.ons.gov.uk/peoplepopulationandcommunity/personalandhouseholdfinances/ incomeandwealth/bulletins/householdincomeinequalityfinancial/financialyearending 2019, accessed April 11, 2020. A darker interpretation is that inequality is higher than the survey and tax data show, because of increasing tax dodging and the use of tax havens among the top 1 percent, see Annette Alstadsæter, Niels Johannesen, and Gabriel Zucman, "Tax Evasion and Inequality." *American Economic Review* 109, no. 6 (2019): 2073–2103.

58. The rising tide quote was popularized by John F. Kennedy who used it in a 1963 speech. The Stiglitz quote is from Joseph E. Stiglitz, "Inequality and Economic Growth," in *Rethinking Capitalism*, ed. Michael Jacobs and Mariana Mazzucato (Hoboken, NJ: Wiley-Blackwell, 2016), 134–155.

59. Sarah Elena Windolph, "Assessing Corporate Sustainability through Ratings: Challenges and Their Causes," *Journal of Environmental Sustainability* 1, no. 1 (November 1, 2011):

1–22; SustainAbility, *Rate the Raters 2019: Expert Views on ESG Ratings* (SustainAbility, February 25, 2019), https://sustainability.com/our-work/reports/rate-raters-2019/.

60. Source for expert's ranking of sustainability leaders: The 2018 GlobeScan-SustainAbility Leaders Survey. https://globescan.com/wp-content/uploads/2018/06/GlobeScan-Sustain Ability-Leaders-Survey-2018-Report.pdf, accessed April 10, 2019. Other leading companies, including Ørsted, are scored at Corprate Knight's Global 100 ranking: https://www .corporateknights.com/reports/2019-global-100/2019-global-100-results-15481153/, accessed April 10, 2020.

61. CDP, from Carbon Disclosure Project, is a leading not-for-profit charity that runs the global disclosure system for investors, companies, cities, states, and regions to manage their environmental impacts, including carbon, forest, and water impact reporting. See www .CDP.net for info on methods and data on particular companies.

62. Source for figure 7.8: author. Data sources: Factiva database (IFRS standard) for financial data, and CDP database, scope 1+2 (market-based), plus annual reports where data was lacking. Inflation adjusted.

63. As in these two sources: https://oilprice.com/Energy/Energy-General/The-Greenest-Oil -Companies-In-The-World.htm, and https://www.rigzone.com/news/oil_gas/a/148169/oil _gas_firms_make_global_100_most_sustainable_corporations_list/.

64. Jesse Jenkins, "Historic Paths to Decarbonization," April 3, 2012, http://thebreakthrough .org/archive/which_nations_have_reduced_car; G. P. Peters et al., "Growth in Emission Transfers via International Trade from 1990 to 2008," *Proceedings of the National Academy of Sciences* 108, no. 21 (May 24, 2011): 8903–8908; Thomas O Wiedmann et al., "The Material Footprint of Nations," *Proceedings of the National Academy of Sciences* 112, no. 20 (2015): 6271–6276.

65. Pierre Friedlingstein et al., "Global Carbon Budget 2019," *Earth System Science Data* 11, no. 4 (December 4, 2019): 1783–1838, section 2.1.3.

66. Jorgen Randers, "Greenhouse Gas Emissions per Unit of Value Added ('GEVA')—A Corporate Guide to Voluntary Climate Action," *Energy Policy* 48 (September 2012): 46–55; Colin Haslam et al., "Accounting for Carbon and Reframing Disclosure: A Business Model Approach," *Accounting Forum* 38, no. 3 (September 2014): 200–211.

67. See https://ghgprotocol.org/about-us, accessed March 20, 2020.

68. William Nordhaus, "Climate Clubs: Overcoming Free-Riding in International Climate Policy," American Economic Review 105, no. 4 (April 2015): 1339–1370.

69. Harald Fuhr, Thomas Hickmann, and Kristine Kern, "The Role of Cities in Multi-Level Climate Governance: Local Climate Policies and the 1.5 °C Target," *Current Opinion in Environmental Sustainability* 30 (February 2018): 1–6, https://doi.org/10.1016/j.cosust .2017.10.006; Luis Gomez Echeverri, "Investing for Rapid Decarbonization in Cities," *Current Opinion in Environmental Sustainability* 30 (February 2018): 42–51, https://doi

.org/10.1016/j.cosust.2018.02.010; William Solecki et al., "City Transformations in a 1.5°C Warmer World," *Nature Climate Change* 8, no. 3 (March 2018): 177–181.

70. Data source for figure 7.9 on Leeds: Andy Gouldson et al., "Exploring the Economic Case for Climate Action in Cities," *Global Environmental Change* 35 (November 2015): 93–105. Data series from figure 1d, emissions per unit of GDP 2000–2025, numbers from supplemental materials, and then inverted from intensity to carbon productivity (CAPRO).

71. Gouldson et al., "Exploring the Economic Case for Climate Action in Cities," 101.

72. Data source for figure 7.10: Euromonitor International, "Income Inequality Ranking of the World's Major Cities." I assume a linear rate of change of 2.2 percent between the two years, as the source does not give annual figures.

73. See Weizsäcker, *Factor Five*, and the Further Reading section for more.

74. Managi and Kumar, *Inclusive Wealth Report 2018*.

PART III

1. Steffen et al., "The Trajectory of the Anthropocene"; IPBES (Intergovernmental Science-Policy Platform on Biodiversity and Ecosystem Services), *Global Assessment of Biodiversity and Ecosystem Services* (IPBES, 2019).

CHAPTER 8

1. John Micklethwait and Adrian Wooldridge, *The Company: A Short History of a Revolutionary Idea* (New York: Modern Library, 2005), xv.

2. John Montgomery, "Benefit Corporations," TEDx, filmed in 2012 at Hult International Business School, https://www.youtube.com/watch?v=NfNF9u7X0GU.

3. Joel Bakan, *The Corporation: The Pathological Pursuit of Profit and Power* (New York: Free Press, 2005), 1.

4. Elkington, *Cannibals with Forks*.

5. John Montgomery, "A Value Shift: From Fear to Love, interview by Leslie Lawton," LinkedIn, June 28, 2019, https://www.linkedin.com/pulse/value-shift-from-fear-love-john-montgomery/.

6. Michael B. Dorff, James Hicks, and Steven Davidoff Solomon, "The Future or Fancy? An Empirical Study of Public Benefit Corporations," *SSRN Electronic Journal* (2019).

7. J. Epstein-Reeves and E. Weinreb, *Pioneers of Sustainability: Lessons from Trailblazers* (Weinreb Group, September 17, 2013).

8. George Ogleby, "Jonathon Porritt: Business Leaders Are Better Placed than MPs to Drive the Green Economy," *Edie,* August 26, 2016, http://www.edie.net/news/7/Jonathon-Porritt-green-business-leaders-will-drive-the-future-green-economy/.

9. In several surveys of more than 1,000 top executives from twenty-seven industries across 103 countries, conducted by Accenture and others, a common finding is that practitioners struggle to integrate sustainability with business strategy: Accenture & UN Global Compact, *The UN Global Compact-Accenture CEO Study on Sustainability 2013: Architects of a Better World* (Accenture, 2013), 1–59; Accenture & UN Global Compact, *Transforming Partnerships for the SDGs* (Accenture, 2018); BSR/Globescan, *State of Sustainable Business Survey* (October 2019); Jenny Davis-Peccoud, "Transforming Business for a Sustainable Economy," Bain, August 1, 2018, https://www.bain.com/insights/transforming-business -for-a-sustainable-economy/.

10. Gareth Kane, *The Green Executive: Corporate Leadership in a Low Carbon Economy* (London: Earthscan, 2011); Heim, *Killing Sustainability*.

11. George Serafeim, Robert Zochowski, and Jen Downing, *Impact-Weighted Financial Accounts* (Harvard Business School, 2019).

12. The six-step stairs model and figure 8.1 were made by the author, but draw on a number of pre-existing models, most importantly from Kane, *The Green Executive*.

13. See https://www.salesforce.org/about-us/, accessed January 4, 2020.

14. https://www.wearestillin.com, accessed January 4, 2020.

15. Gordon L. Clark, Andreas Feiner, and Michael Viehs, "From the Stockholder to the Stakeholder," *SSRN Electronic Journal* (2015).

16. John Brady, Alison Ebbage, and Ruth Lunn, eds., *Environmental Management in Organizations: The IEMA Handbook*, 2nd ed. (London: Earthscan, 2011).

17. Magali A. Delmas and Sanja Pekovic, "Environmental Standards and Labor Productivity: Understanding the Mechanisms That Sustain Sustainability," *Journal of Organizational Behavior* 34, no. 2 (February 2013): 230–252, 245; Magali Delmas and Sanja Pekovic, "The Engaged Organization: Human Capital, Social Capital, Green Capital and Labor Productivity," *Academy of Management Proceedings, Academy of Management* (2013): 10483.

18. Paul Palmer, *Getting to Zero Waste:Universal Recycling as a Practical Alternative to Endless Attempts to "Clean Up Pollution"* (Sebastopol, CA: Purple Sky Press, 2004).

19. George Marshall, *Don't Even Think about It: Why Our Brains Are Wired to Ignore Climate Change* (New York: Bloomsbury, 2014), 205.

20. One status update on Walmart's wage policy is given here: https://www.bizjournals.com/ twincities/news/2019/06/06/walmart-calls-for-increased-minimum-wage.html, accessed March 1, 2020.

21. See https://www.ikea.com/ms/en_KR/this-is-ikea/people-and-planet/energy-and-resources, accessed January 19, 2019.

22. Source: Philips, http://www.philips.com/a-w/about/company/suppliers/supplier-sustainability/ our-programs/supplier-sustainability-assessment.html, accessed July 2, 2020.

23. Mark Pagell and Zhaohui Wu, "Business Implications of Sustainability Practices in Supply Chains," in *Sustainable Supply Chains*, ed. Yann Bouchery et al., vol. 4 (Cham: Springer International Publishing, 2017), 339–53.

24. McDonough, *The Upcycle*.

25. See Lovins, "How Big Is the Energy Efficiency Resource?"

26. Capgemini, The Climate Group, and CDP, *Making Business Sense: How RE100 Companies Have an Edge on Their Peers—Insights Report* (London, September 2018).

27. Source: WEForum, https://www.weforum.org/agenda/2019/09/owner-ikea-exceed-renewable-energy-goal-years-end/, accessed April 10, 2020.

28. See Lovins, "How Big Is the Energy Efficiency Resource?" and the Further Reading section.

29. Source: EnergiOgKlima.no, https://energiogklima.no/nyhet/tizir-sikter-pa-gront-hydrogen/, accessed March 1, 2020.

30. Duncan Kushnir et al., "Adopting Hydrogen Direct Reduction for the Swedish Steel Industry," *Journal of Cleaner Production* 242 (January 2020): 118185.

31. An overview of such initiatives is given in the Further Reading section.

32. The 2013 survey found 93 percent. Accenture & UN Global Compact, "The UN Global Compact-Accenture CEO Study on Sustainability 2013: Architects of a Better World"; in the 2016 survey it increased to 97 percent. See https://www.accenture.com/us-en/insight-un-global-compact-ceo-study, accessed April 10, 2020.

33. A finding repeated in the BSR Globescan study, *State of Sustainable Business Survey* (October 2013).

34. Accenture & UN Global Compact, "The UN Global Compact-Accenture CEO Study on Sustainability 2013: Architects of a Better World," 23.

35. Clark, Feiner, and Viehs, "From the Stockholder to the Stakeholder"; Robert G. Eccles, Ioannis Ioannou, and George Serafeim, "The Impact of Corporate Sustainability on Organizational Processes and Performance," *Management Science* 60, no. 11 (November 2014): 2835–2857; Bob Willard, *The New Sustainability Advantage: Seven Business Case Benefits of a Triple Bottom Line*, rev. ed. (Gabriola Island, BC: New Society Publishers, 2012); Gunnar Friede, Timo Busch, and Alexander Bassen, "ESG and Financial Performance: Aggregated Evidence from More than 2000 Empirical Studies," *Journal of Sustainable Finance & Investment* 5, no. 4 (October 2, 2015): 210–233.

36. Accenture reports that, in 2019, one in three CEOs of the world's largest companies cite "lack of market pull" as a top barrier to sustainable business. Further, political uncertainty is reducing or stalling their sustainability efforts for 42 percent of CEOs while a third (34 percent) specify market closures and limitations on free trade as hindrances. Over half (55 percent) of CEOs say that they are facing a key trade-off: pressure to operate under extreme cost-consciousness versus investing in longer-term strategic objectives that are at the heart of sustainability. See Accenture & UN Global Compact, "The UN Global

Compact-Accenture CEO Study on Sustainability 2013: Architects of a Better World," 21; Accenture & UN Global Compact, "The Decade to Deliver: The CEO Study Program," 2019, 14.

CHAPTER 9

1. Stoknes, *What We Think about When We Try Not to Think about Global Warming*, ch. 8–9.

2. Source for figure 9.1: BSR/Globescan, *State of Sustainable Business Survey* (2015).

3. Nielsen company, "Sustainable Shoppers Buy the Change They Wish to See in the World," 2018.

4. Anne Oudersluys, "5 Insights Every Brand Should Know about the Sustainable Shopper, from Nielsen's Latest Sales Data," Core Impact—Social Impact & Purpose Driven Marketing, February 2019, https://www.coreimpactstrategy.com/blog1/5-insights-on-sustainable -consumer, accessed April 20, 2020; Randi Kronthal-Sacco et al., "Sustainable Purchasing Patterns and Consumer Responsiveness to Sustainability Marketing," *SSRN 3465669* (August 2019).

5. Paul C. Stern, "New Environmental Theories: Toward a Coherent Theory of Environmentally Significant Behavior," *Journal of Social Issues* 56, no. 3 (January 2000): 407–424; Katherine White, Rishad Habib, and David J. Hardisty, "How to SHIFT Consumer Behaviors to Be More Sustainable: A Literature Review and Guiding Framework," *Journal of Marketing* 83, no. 3 (May 2019): 22–49.

6. Brian J. Fogg, "A Behavior Model for Persuasive Design," *Proceedings of the 4th international Conference on Persuasive Technology* (ACM, 2009), 40.

7. Charles Duhigg, *Power of Habit: Why We Do What We Do in Life and Business* (New York: Random House, 2014).

8. Stern, "New Environmental Theories"; Tom Crompton, *Common Cause: The Case for Working with Our Cultural Values* (World Wildlife Fund UK, September 2010).

9. Stoknes, *What We Think About*, ch. 11; Cass R. Sunstein, "Behavioral Economics, Consumption, and Environmental Protection," *SSRN Electronic Journal* (2013); Richard H. Thaler and Cass R. Sunstein, *Nudge: Improving Decisions about Health, Wealth, and Happiness* (New Haven: Yale University Press, 2008).

10. WWF, CDP, and McKinsey & Company, "The 3% Solution Driving Profits Through Carbon Reductions"; Heim, *Killing Sustainability*.

11. Accenture & UN Global Compact, "The Decade to Deliver," figure 3, p. 25.

12. Camila Huerta Alvarez, Julius Alexander McGee, and Richard York, "Is Labor Green?," *Nature and Culture* 14, no. 1 (2019): 17–38.

13. Marco Casini, "Active Dynamic Windows for Buildings: A Review," *Renewable Energy* 119 (2018): 923–934.

14. See, for instance, Michael Shuman, *The Local Economy Solution* (Vermont: Chelsea Green, 2015), or the Global Citizen blog: https://www.globalcitizen.org/en/content/impact -investment-what-is-it-how-to-make-money/, accessed April 10. 2020.

15. See Kiva.org, https://www.kiva.org/about/due-diligence/risk, accessed April 10, 2020.

16. Gilens and Page, "Testing Theories of American Politics"; Christopher H. Achen and Larry M. Bartels, *Democracy for Realists: Why Elections Do Not Produce Responsive Government* (Princeton, NJ: Princeton University Press, 2017).

17. Anthony Fowler and Michele Margolis, "The Political Consequences of Uninformed Voters," *Electoral Studies* 34 (2014): 100–110.

18. Russell J. Dalton, *The Good Citizen: How a Younger Generation Is Reshaping American Politics*, Second edition (Thousand Oaks, CA: CQ Press, 2016), loc 263.

19. See https://citizensclimatelobby.org/energy-innovation-and-carbon-dividend-act/, accessed January 10, 2020.

20. James Hillman, *City & Soul*, Uniform Edition of the Writings of James Hillman, v. 2 (Putnam, CT: Spring Publications, Inc., 2006), loc. 1851.

21. Paul Hawken, *Blessed Unrest: How the Largest Social Movement in History Is Restoring Grace, Justice and Beauty to the World.* (New York: Penguin, 2007), loc. 134.

22. Arild Vatn and Daniel W Bromley, "Choices without Prices without Apologies," *Journal of Environmental Economics and Management* 26, no. 2 (1994): 129–148; Arild Vatn, *Institutions and the Environment* (Cheltenham, UK: Edward Elgar Publishing, 2007).

CHAPTER 10

1. Lakoff, *The Political Mind*; Westen, *The Political Brain*.

2. GreenBiz Group and TruCost ESG, "State of Green Business 2019."

3. A. Gustafson et al., *A Growing Majority of Americans Think Global Warming Is Happening and Are Worried, Yale Program on Climate Change Communication* (New Haven, CT: Yale University and George Mason University, 2019); Matthew T. Ballew et al., "Climate Change in the American Mind: Data, Tools, and Trends," *Environment: Science and Policy for Sustainable Development* 61, no. 3 (May 4, 2019): 4–18.

4. Mark Reed et al., "What Is Social Learning?," *Ecology and Society* 15, no. 4 (2010).

5. Ortony, *Metaphor and Thought*; Lakoff and Johnson, *Metaphors We Live By*.

6. Andrew J. Hoffman, "Talking Past Each Other? Cultural Framing of Skeptical and Convinced Logics in the Climate Change Debate," *Organization & Environment* 24, no. 1 (March 2011): 3–33; David Takacs, "Beyond Zero-Sum Environmentalism," *Environmental Law Reporter News & Analysis* 47 (2017): 10328.

7. Daniel Stedman Jones, *Masters of the Universe: Hayek, Friedman, and the Birth of Neoliberal Politics* (Princeton, NJ: Princeton University Press, 2014).

8.   Paul Krugman, "The Tax Cut Con," *New York Times*, September 14, 2003, http://www
     .nytimes.com/2003/09/14/magazine/the-tax-cut-con.html.

9.   Jones, *Masters of the Universe*; Philip Mirowski, Dieter Plehwe, and Askews and Holts
     Library Services, *The Road from Mont Pelerin: The Making of the Neoliberal Thought Collec-
     tive* (Cambridge, MA: Harvard University Press, 2015).

10. Milton Friedman, "What Every American Wants," *Wall Street Journal,* January 15, 2003,
     http://www.wsj.com/articles/SB1042593796704188064.

11. Source for figure 10.1: Pew Research Trust, https://www.people-press.org/2019/04/11/
     public-trust-in-government-1958-2019/.

12. Source: Pew Research Trust, https://www.people-press.org/2019/04/11/public-trust-in
     -government-1958-2019/.

13. Gallup, "Americans Still See Big Government as Top Threat," Gallup.com, January 5, 2017,
     http://www.gallup.com/poll/201629/americans-big-government-top-threat.aspx.

14. Source for figure 10.2: OECD, *Government at a Glance 2019* (Paris: OECD Publishing,
     2019), https://doi.org/10.1787/gov_glance-2017-en, accessed November 10, 2019.

15. As in this quote: "the frustration with Washington [is] the sense that the system is broken.
     Voters feel that we have no control and that government has gone wild. Even people who
     don't watch the news or closely follow politics are aware of the 'overreach.'" From Marita
     Noon, "Rolling Back the Tide of Big Government Overreach," *Heartland,* October 10, 2015,
     http://blog.heartland.org/2015/10/rolling-back-the-tide-of-big-government-overreach/.
     More examples: Kevin Kossar and Phillip Wallach, "Stopping Big Government: Why We
     Need a Congressional Regulation Office," Fox News, November 16, 2016, http://www
     .foxnews.com/opinion/2016/11/16/stopping-big-government-why-need-congressional
     -regulation-office.html; Ron Arnold, "Big Government, Foundations Undermine Scientific
     Integrity," Heartland Institute, April 6, 2015, https://www.heartland.org/news-opinion/
     news/big-government-foundations-undermine-scientific-integrity; or Jeff Poor, "Lim-
     baugh: Big Government Liberalism Was Told to Go to Hell," *Breitbart*, November 9, 2016,
     http://www.breitbart.com/video/2016/11/09/limbaugh-big-government-liberalism-told
     -go-hell/. All accessed April 10, 2020.

16. John Komlos, "Reaganomics: A Watershed Moment on the Road to Trumpism," *Economists'
     Voice* 16, no. 1 (January 5, 2019).

17. Charles G. Koch, *Good Profit: How Creating Value for Others Built One of the World's Most
     Successful Companies* (New York: Crown Business, 2015).

18. Mariana Mazzucato, *The Entrepreneurial State: Debunking Public vs. Private Sector Myths*
     (New York: PublicAffairs, 2015), loc. 1101.

19. George Lakoff, *The All-New Don't Think of an Elephant! Know Your Values and Frame the
     Debate* (White River Junction, VT: Chelsea Green Publishing, 2014), loc. 958; my italics.

20. Mazzucato, *The Entrepreneurial State*, 109.

21. Alex Trembath et al., "Where the Shale Gas Revolution Came From," *Breakthrough Institute* 23 (2012).

22. Mazzucato, *The Entrepreneurial State,* loc. 552.

23. Mazzucato, *The Entrepreneurial State,* loc. 585.

24. Mazzucato, *The Entrepreneurial State,* loc. 646; my italics.

25. Andrew Winston, *The Big Pivot: Radically Practical Strategies for a Hotter, Scarcer, and More Open World* (Cambridge, MA: Harvard Business Review Press, 2014), loc. 328.

26. OECD, Public Procurement for Sustainable and Inclusive Growth (Paris: OECD Publishing, 2016), http://www.oecd.org/gov/ethics/Public-Procurement-for%20Sustainable -and-Inclusive-Growth_Brochure.pdf.

27. Luis Mundaca and Jessika Luth Richter, "Assessing 'Green Energy Economy' Stimulus Packages: Evidence from the U.S. Programs Targeting Renewable Energy," *Renewable and Sustainable Energy Reviews* 42 (February 2015): 1174–1186.

28. In an age where quantitative easing has been "used up" as a means to stimulate the economy, particularly by giving fiscal transfers to low-income groups. See discussion in Maarten van Rooij and Jakob de Haan, "Would Helicopter Money Be Spent? New Evidence for the Netherlands," *Applied Economics* (May 7, 2019): 1–19.

29. Skjelvik, Ibenholt, and Bruvoll, *Greening the Economy,* 97.

30. Peter Barnes, *Capitalism 3.0: A Guide to Reclaiming the Commons* (Readhowyouwant.com Ltd, 2014).

31. M. Jakob and O. Edenhofer, "Green Growth, Degrowth, and the Commons," *Oxford Review of Economic Policy* 30, no. 3 (2014): 447–468.

32. Eleanor Roy, "New Zealand River Granted Same Legal Rights as Human Being," *Guardian*, March 16, 2017, https://www.theguardian.com/world/2017/mar/16/new-zealand -river-granted-same-legal-rights-as-human-being.

33. Elinor Ostrom, "Revisiting the Commons: Local Lessons, Global Challenges," *Science* 284, no. 5412 (1999): 278–282, https://doi.org/10.1126/science.284.5412.278; Elinor Ostrom, *Governing the Commons: The Evolution of Institutions for Collective Action* (Cambridge: Cambridge University Press, 2015).

34. European Bank for Reconstructioun and Development, *Transition Report 2019–20 Better Governance, Better Economies* (London: EBRD, 2019).

35. Bo Rothstein and Dietlind Stolle, "The State and Social Capital: An Institutional Theory of Generalized Trust," *Comparative Politics* 40, no. 4 (2008): 441–459.

CHAPTER 11

1. Stuchtey, Enkvist, and Zumwinkel, *A Good Disruption*, loc. 896.

2. GreenBiz Group and TruCost ESG, *State of Green Business 2020*, 3.

3. See "What Is Impact?," Impact Management, https://impactmanagementproject.com/impact-management/what-is-impact/, accessed March 1, 2020.

4. For dematerialization, see McAfee, *More from Less*; for CO_2, see Le Quéré et al., "Drivers of Declining CO2 Emissions in 18 Developed Economies"; for trade, see Wood et al., "Beyond Peak Emission Transfers."

5. Clark, Feiner, and Viehs, "From the Stockholder to the Stakeholder"; Ram Nidumolu, PJ Simmons, and Terry F. Yosie, "Sustainability and the CFO Challenges, Opportunities and Next Practices," *Corporate EcoForum and World Environment Center* (April 2015); Marc J. Epstein and Adriana Rejc, *Making Sustainability Work*, 2nd ed. (San Francisco: Greenleaf Publishing, 2014).

6. Rob Terwel, John Kerkhoven, and Frans W. Saris, "Carbon Neutral Aviation," *Europhysics News* 50, nos. 5–6 (September 2019): 29–32; Energy Transitions Commission, *Mission Possible*, ch. 3.

7. On Blackrock, see Larry Fink's statement here: Andrew Sorkin, "BlackRock CEO Larry Fink: Climate Crisis Will Reshape Finance," *New York Times,* January 14, 2020, https://www.nytimes.com/2020/01/14/business/dealbook/larry-fink-blackrock-climate-change.html.

8. An overview over solutions, examples and analysis are described in the Further Reading section.

9. Randers et al., *Transformation Is Feasible*.

10. Earth3 is a work in progress. Models and data can be downloaded from www.2052 .info/earth3. The team is researching a fuller systems-dynamics, endogenous model that also incorporates more of the financial and fair (i.e., interest rates and worker-owner dimensions). Our modeling approach with data sources are is described in Randers et al., *Achieving the 17 Sustainable Development Goals within 9 Planetary Boundaries*; Collste et al., "The Empirical Bases for the Earth3 Model"; Jorgen Randers et al., "A User-Friendly Earth System Model of Low Complexity: The ESCIMO System Dynamics Model of Global Warming towards 2100," *Earth System Dynamics* 7, no. 4 (2016): 831–850.

11. One such central database is in Robert C. Feenstra, Robert Inklaar, and Marcel P. Timmer, "The Next Generation of the Penn World Table," *American Economic Review* 105, no. 10 (2015): 3150–3182, available at www.ggdc.net/pwt. For more on the databases, see Collste et al., "The Empirical Bases for the Earth3 Model."

12. We grade the SDG achievement in a simple way: An achieved goal (green) means 1 point. A goal that has passed the halfway target is 0.5 point (yellow). An unachieved goal (red) is 0 points. See table 5.5 in appendix 1 of the *Transformation Is Feasible* report for details on goals, chosen indicators, and thresholds. Targets and thresholds are based on Sachs et al., *SDG Index & Dashboards: A Global Report*.

13. The safety margin is calculated similarly to SDG score: each PB has a red high-risk zone (0), yellow mid-state zone (0.5), and a green safe zone (1). We do not give different weight to the PBs.

14. Source for figures 11.1–11.4: Randers et al., *Transformation Is Feasible*.

15. On challenge-and-response, along with the classic Toynbee theory, see Bruce Alberts, *Sustainable Development: The Challenge of Transition*, vol. 2 (Cambridge: Cambridge University Press, 2000), and the concept of plot in Peter Schwartz, *The Art of the Long View: Paths to Strategic Insight for Yourself and Your Company* (New York: Bantam Doubleday, 1996).

16. IEA/OECD, *World Energy Investment 2018*, https://www.iea.org/wei2018/, accessed August 1, 2018.

17. "Levelized Cost of Energy and Levelized Cost of Storage 2019," Lazard.com, November 7, 2019, https://www.lazard.com/perspective/lcoe2019.

18. J. Rockström, O. Gaffney, J. Rogelj, M. Meinshausen, N. Nakicenovic, and H. J. Schellnhuber, "A Roadmap for Rapid Decarbonization," *Science* 355, no. 6331 (2017): 1269–1271.

19. See Hawken, *Drawdown*, 39; Willett et al., "Food in the Anthropocene: The EAT–Lancet Commission on Healthy Diets from Sustainable Food Systems."

20. D. Bossio et al., "The Role of Soil Carbon in Natural Climate Solutions," *Nature Sustainability*, March 16, 2020.

21. More on what we mean with the "China model" in the *Transformation Is Feasible*, box 2; Daniel Bell, *The China Model: Political Meritocracy and the Limits of Democracy* (Princeton, NJ: Princeton University Press, 2015); Yuen Yuen Ang, "The Real China Model," *Foreign Affairs*, July 10, 2018, https://www.foreignaffairs.com/articles/asia/2018-06-29/real-china-model.

22. Erik S. Reinert, *How Rich Countries Got Rich . . . and Why Poor Countries Stay Poor* (London: Constable, 2008).

23. Olav Lundstøl, "Tax in Development: Towards a Strategic Aid Approach" (International Center for Tax and Development, 2018).

24. OECD, *In It Together*; IMF, *Fostering Inclusive Growth: G-20 Leaders' Summit*; Loungani, "The Power of Two."

25. Sue Duke, "The Key to Closing the Gender Gap? Putting More Women in Charge," World Economic Forum, November 2, 2017, https://www.weforum.org/agenda/2017/11/women-leaders-key-to-workplace-equality.

26. Business and Sustainable Development Commission, *Better Business Better World*, 27.

27. Bateman, *The Sex Factor*; Marcus Noland and Tyler Moran, "Study: FIRMS with More Women in the C-Suite Are More Profitable," *Harvard Business Review*, February 8, 2016.

28. See Randers et al., *Transformation Is Feasible*. See also the rapid social development SSP1 scenario in Wolfgang Lutz et al., *Demographic and Human Capital Scenarios for the 21st Century: 2018 Assessment for 201 Countries* (Publications Office of the European Union, 2018), 8.

29. Weber, *Enlivenment*, 17; my italics.

Further Reading

The full bibliography for this volume is available online at http://www .growthcompass.info.

Here is an annotated select list of key (a) overview projects, (b) books, and (c) reports related to healthy growth.

(A) HEALTHY GROWTH SOLUTIONS: OVERVIEWS AND CASE COMPILATIONS

Drawdown is both a beautiful, bestselling book and an active, developing project online at www.drawdown.org. It amounts to an inspiring treasure chest of one hundred climate solutions and upcoming attractions that can, taken together, reverse global warming. Author Paul Hawken is following up with a successor on solutions for not just climate mitigation but also *Regeneration*, in 2021.

Global Opportunity Explorer and Sustainia 100. The Danish NGO Sustainia started scanning and screening sustainability innovations many years ago and published annual reports of the top one hundred examples. Now, along with UN Global Compact and auditing company DNV-GL, they maintain a database containing hundreds of healthy-growth solutions on www.GOExplorer.org.

Mission Innovation is a global initiative of twenty-four countries and the European Commission (on behalf of the European Union). They have presented no fewer than a thousand clean energy cases and solutions with the potential to deliver close to three gigatons of avoided emissions by 2030. See www.misolutionframework.net/Innovations.

Climate-KIC is a European Knowledge and Innovation Community, working toward prosperous, inclusive, climate-resilient societies founded on a circular, zero-carbon economy. More than 1,600 climate-positive companies have been incubated, they report. See www.climate-kic.org.

GreenBiz is a large US media and events company, founded by thought-leader Joel Makower, that advances the opportunities at the intersection of business, technology, and sustainability. It reports on leadership, cases, analysis, and corporate news in aligning environmental responsibility with profitable business practices. See www.greenbiz.com and the company's series of *Verge* conferences.

The *Ellen MacArthur Foundation* works in education, business, government, analysis, systemic initiatives, and communications to accelerate the transition to a circular economy, driving radical resource productivity. They maintain a large database of circular economy business cases. See www.ellenmacarthurfoundation.org/case-studies.

The Blue Economy is an initiative founded by clean-tech entrepreneur Gunther Pauli, in order to highlight more than one hundred creative economic solutions to the current challenges. It also became a report to the Club of Rome. See www.theblueeconomy.org.

Cradle-to-Cradle is an initiative founded by designers and authors McDonough and Braungarten. It's developed into a certification process. A database with more than two hundred cases is available at www.c2c-centre.com/companies-and-organizations.

The *Green Growth Knowledge Platform* is a global network of experts and organizations dedicated to providing the policy, business, and finance communities with knowledge, guidance, data, and tools to transition to an inclusive green economy. Supported by the OECD, UNEP, the World Bank, and others, they offer quick and easy access to the latest research, case studies, toolkits, learning products. See www.greengrowthknowledge.org.

(B) BOOKS

I mention two classics plus a select handful of newer works that I view as being largely in alliance with this volume, deepening the analysis here into various closely related domains.

The first is Ernst von Weizsäcker's update to his classic *Factor Four*, aptly named *Factor Five: Transforming the Global Economy through 80% Improvements in Resource Productivity* (London: Earthscan, 2009). Radical resource productivity is often confused with gradual, diminishing-return incremental efficiency improvements, which often only lead to gray growth. These two books convincingly lay out the possibilities and strategies for radical resource productivity and genuine green growth.

The other classic is Hawken, Lovins, and Lovins's *Natural Capitalism*. This book blew my mind when it appeared in 1999. Even if it is been out a long time, and the examples and numbers are now dated, most of the principles and strategies remain valid today. Paul Hawken, Amory Lovins, and L. Hunter Lovins, *Natural Capitalism: Creating the Next Industrial Revolution* (Boston: Little, Brown, 1999).

Among newer titles, we find *A Good Disruption* by Stuchtey, Enkvist, and Zumwinkel. This is probably the best book on the subject I've read. Its main argument is the need for systemic and circular—not just incremental and cost-cutting—innovations in order to redefine growth. Then, the digital technologies can give not just a disruption, but can contribute to a good disruption. Martin Stuchtey, Martin R., Per-Anders Enkvist, and Klaus Zumwinkel, *A Good Disruption: Redefining Growth in the Twenty-First Century* (London: Bloomsbury, 2016).

Andrew Winston argues that what has until now been called green business, or sustainability, cannot be a side department or a niche conversation in commerce. He goes into the depth of corporate strategy, outlining how every business must pivot so that solving the world's biggest challenges profitably becomes the core pursuit of business. Andrew Winston, *The Big Pivot: Radically Practical Strategies for a Hotter, Scarcer, and More Open World* (Cambridge, MA: Harvard Business Review Press, 2014).

In *Doughnut Economics*, Kate Raworth gives a brilliant analysis of why the seven commonly touted "rules" of conventional economic thought don't hold true in the twenty-first century. But in her chapter on growth, she stops short of fully engaging the thorny growth issue, landing briefly on an "agnostic" stance and leaving the door open for a deeper discussion of growth—as this volume has contributed to. Kate Raworth, *Doughnut Economics: Seven Ways to Think like a 21st Century Economist* (White River Junction, VT: Chelsea Green Publishing, 2017).

L. Hunter Lovins's *A Finer Future* retells in an enthusiastic voice the great turning that our societies are, can, and will take in the coming decades. In an inspirational, inclusive way, she and coauthors give experiences, images, and new ideas on how to create a finer future going forward. L. Hunter Lovins, Stewart Wallis, Anders Wijkman, and John Fullerton, *A Finer Future: Creating an Economy in Service to Life* (Gabriola Island, BC: New Society Publishers, 2018).

John Elkington's *Green Swans* sums up his decades-long work in sustainability, taking off from the concepts of Nassim Taleb's *Black Swan*. He highlights why now is the time when unforeseen, exponential growth in green solutions can switch from gradual change to sudden domination of the economic growth arena. The book deepens the outline of the sixth wave, given in chapter 3 this volume. John Elkington, *Green Swans: The Coming Boom in Regenerative Capitalism* (New York: Fast Company, 2020).

(C) SOME RECENT BREAKTHROUGH REPORTS

New Climate Economy Report: Sir Nick Stern is a world-famous economist, who really placed climate economics into the mainstream. Since his massive and pivotal "Stern Review" of 2006, a larger organization has taken the analysis much further than the initial analysis. A decade of expanding, rewriting, and improving has taken the latest version of the report to an inclusive and encompassing overview over the swerve to healthier growth: Global Commission on the Economy and Climate, *Unlocking the Inclusive Growth Story of the 21st Century*, NCER, www.newclimateeconomy.report/.

The World Inequality Report, headed by Facundo Alvadero and cofounded with Thomas Piketty, gives an excellent, unprecedented overview of the current inequality trends. It also discusses how to tackle them and possible futures for a healthier global income inequality: www.wid.world.

The Global 100% Renewables Study: You've probably repeatedly heard many versions of the argument that "renewables cannot power our economy." One surprisingly thorough study shows exactly how it can—and at no extra cost relative to the fossil system. It was developed by eighteen researchers based in Germany and Finland working for four years with detailed modeling the shift to 100 percent renewables, and then writing this impressive report: www.energywatchgroup.org/new-study-global-energy-system-based-100-renewable-energy.

Mission Possible: If 100 percent renewable power is possible, what about the harder-to-abate sectors such as steel, concrete, heavy transport, or aluminum? The international Energy Transition Commission treats these thorny issues in-depth in their report *Mission Possible: Reaching Net-Zero Carbon Emissions from Harder-to-Abate Sectors by Mid-Century* (2018), www.energy-transitions.org/sites/default/files/ETC_MissionPossible _FullReport.pdf.

Exponential Climate Action Roadmap: Building on Johan Rockström et al.'s "Carbon Law" concept, this climate science-based report highlights the most impactful thirty-six solutions that can rolled out during the 2020s, to achieve a near 50 percent accelerated reduction in emissions by 2030: www.exponentialroadmap.org.

A Low Energy Demand (LED) Scenario: Led by Arnulf Grubler at the IIASA in Austria, this breakthrough scientific study explores how radical end-user energy efficiency, integrated with digital and sharing economy innovations, can give better modern lives with much, much less waste by 2050. It is thoroughly modeled with so-called quantitative Integrated Assessment Models (IAMs), which helps consistency and credibility. Both the full article in *Nature Energy* and the detailed supplemental materials are available here: http://pure.iiasa.ac.at/id/eprint/15301/.

All of the above URLs accessed May 3, 2020.

Index

Page numbers followed by an "f" or "t" indicate figures and tables, respectively.